Notes and Comments on VERTEBRATE PALEONTOLOGY

Alfred Sherwood Romer

Notes and Comments on VERTEBRATE PALEONTOLOGY

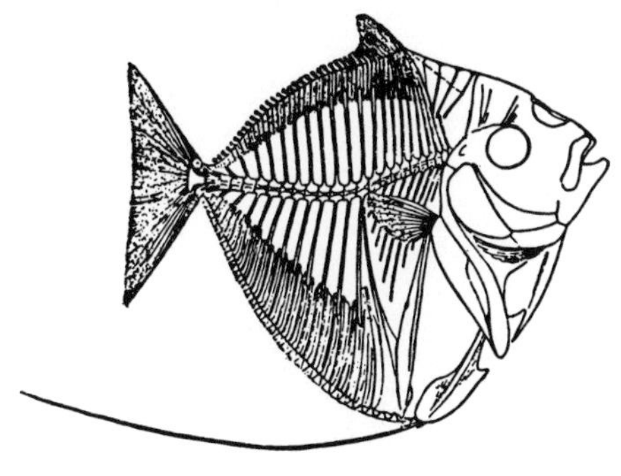

THE UNIVERSITY OF CHICAGO PRESS

CHICAGO AND LONDON

Library of Congress Catalog Card Number: 68–16999

THE UNIVERSITY OF CHICAGO PRESS, CHICAGO 60637
The University of Chicago Press, Ltd., London W.C.1

Published 1968

Printed in the United States of America

Preface

A glance at the Contents of this book might lead a potential reader to assume that there is here a connected account of the phylogeny and classification of fossil vertebrates. This is not at all the case.

During the past few years much of my time has been occupied in preparing a new edition of my *Vertebrate Paleontology*. This has necessitated reading the greater bulk of the literature of the subject for the past twenty years, assessing the advances, some of them very notable, made in various fields, and deciding which point of view to accept for primary text treatment in numerous cases where differing opinions have been advanced on problems of phylogeny and relationship. Since I attempted to keep the length of my text within reasonable bounds and since, as a matter of policy in an elementary work of this sort, I avoided polemics and personalities in the text, little of the results of these studies is to be found in the *Vertebrate Paleontology*. I have, however, thought that it might be of some value to give a further discussion of some of the areas of interest encountered and of the more controversial problems concerned.

Such discussions form the content of the present little volume. Some areas appear to me to contain little that needs discussing, and these are either given short shrift or omitted entirely; others are treated in some detail. The book is thus, as the title indicates, a series of discrete notes and comments on a variety of topics. I have arranged them seriatim in a succession of chapters corresponding to those in the *Vertebrate Paleontology*, but they do not form a connected sequence.

I am indebted here for advice and discussion to many of the colleagues who are listed in the *Vertebrate Paleontology*. In addition, Dr. Richard Estes, Dr. Ernest E. Williams, and Professor Bryan Patterson have read and commented on much of the present text.

A fraction of the publications mentioned are listed in the Bibliography included in *Vertebrate Paleontology* and are referred to by number.

Contents

1

Introductory

Evolutionary theory.—Although some of the most distinguished vertebrate paleontologists of early days—Cuvier, Owen, Agassiz—never came to believe in evolution, most scientists in the field became rapidly converted in post-Darwinian times and, indeed, it is difficult to see how one could work effectively today without an evolutionary background. But until relatively recently there was little concensus of opinion among paleontologists about the principles underlying the evolutionary events which their work was revealing. As with Darwin himself, the trend among paleontologists was strongly toward Lamarckian explanations. The many gaps which existed in phyletic lines (and still exist in many cases) made a belief in evolution by marked "saltations" credible; the seeming tendency (emphasis on "seeming") for many evolutionary series to follow a straightforward, nondeviating course led many toward or to a belief in orthogenesis, with mystical connotations of a directing force. Several distinguished paleontologists developed their own special theories—Cope his "kinetogenesis" first cousin to Lamarckian ideas; the aristocratic Osborn his theory of aristogenesis. Dollo's law of the irreversibility of evolution was overworked; this was properly restricted by its author to the thesis that once a structure was completely eradicated from the hereditary potentialities of an animal it would not redevelop. Later authors, however, tended to extend this "law" to include the absurd idea that structures once gained could not be lost.[1]

Numbered references in the text are to the Bibliography in my *Vertebrate Paleontology*, 3d ed. (Chicago: University of Chicago Press, 1966).

[1] For an example of this, see the phyletic table of the Proboscidea in Osborn's monograph (580). In almost no case is any proboscidean thought to be descended from any known form of earlier times—nearly every form otherwise a possible ancestor of another is debarred from an ancestral position because of the assumption that characters present in the older type but absent in the later form could not be eliminated.

Similar uncertainties and multiplicities of theories were, of course, present among students of Recent animals as well as among paleontologists in the post-Darwinian decades, due primarily to the fact that there was almost complete ignorance about the possible mechanisms of heredity and genetic change. It was not until the present century was well under way that the science of genetics had progressed sufficiently to give us the necessary basic knowledge of heredity. Morgan and his school, to whom we are principally indebted for these advances, were not primarily concerned with evolutionary aspects, but during, particularly, the 1930's there developed, in such men as Fisher, Haldane, and Wright, a school of population geneticists who studied the possible applications of our newly found genetic mechanisms to evolutionary events.

The 1940's saw the initiation of a major development in the field of evolutionary theory. A chance conversation between Professor Walter Bucher of Columbia, then in charge of the earth science section of the National Research Council, and Professor Glenn L. Jepsen of Princeton led to the suggestion that a conference of geneticists and paleontologists, bringing together those interested in theory and in paleontological factual data, might be useful. Two preliminary conferences of this sort (to which came also systematists interested in evolution) were held with obvious profit; continuation of this attempt at synthesis of interests led shortly after the war to holding at Princeton a major conference on evolutionary theory and, presently, to the establishment of a society for the study of evolution and of a successful journal.

Even in the preliminary conferences it became apparent that despite the divergencies of thought among paleontologists, the geneticists and vertebrate paleontologists were, on the whole, not too far apart; in great measure the actual evolutionary events seen in the fossil record could be reasonably interpreted in the light of theories based on genetic knowledge. Today, I think, the great majority of those interested in evolutionary processes, whether as geneticists, systematists, or vertebrate paleontologists, are in essential agreement about the underlying principles, and a modern type of evolutionary theory has emerged. Much of it rests on the fundamental Darwinian thesis of natural selection, and hence it is sometimes referred to as neo-Darwinism.

But this term has been used in the past in variable fashion, and so much has been added theoretically from genetics and from factual data in paleontology and systematics that it is probably better to refer to our modern concept, as Simpson particularly has urged, as a new, synthetic theory of evolution. This modern point of view on evolutionary processes is given in every recent text dealing with the subject; Simpson (1953, etc.) has especially well expounded the fashion in which theory has expressed itself in the actual evolutionary story.

Parallelism and polyphylety.—The evolutionary phenomenon which causes the greatest difficulty in the attempt to work out phylogenies is that of parallelism. In the early days of evolutionary studies it was fondly assumed that the characters which typify a group of animals—whether genus, family, or higher category—had to be inherited by its members from a common ancestor. This we now know is not, unfortunately, the case in many instances. Time after time, we find that given features of the members of a group appear to have been independently developed, in parallel fashion, and were not present in a common ancestor. This is, of course, to be expected from the known facts of genetics. Given two related forms with a similar genetic constitution, it is reasonable that similar viable mutations might occur in each, and if the two are living under similar conditions, selection would tend to operate similarly on the available mutants.

Parallelism is, then, surely known to have occurred among closely related forms, and may be suspected in many other cases. Further, with this background of parallelism in lower taxonomic categories, one may reasonably suspect that parallelism on a broad scale may have operated at the order or class level. Currently the advocacy of polyphyletic origin of major groups is fashionable. In some instances such suggestions have been based upon the fact that group distinctions are arbitrary, and with increasing knowledge of fossil forms close to the boundary, it seems probable that the dividing line may have been crossed independently by several related types. For example, this is not improbably the case as regards the older Mesozoic mammals. Here we are dealing with a limited type of parallelism. But before advocating a claim that a major group has been derived from two or more ancestral types which were far removed from one another phylogenetically, serious thought should be given to the matter, and the proposal should be based not upon one or a few skeletal features, but upon the total organization—including embryology and "soft anatomy," as well as skeletal features. For instance, it has several times been proposed that reptiles were diphyletic in origin, derived from two very different amphibian groups. But when we consider the complex type of amniote development—the shelled and large-yolked egg, the growth of specific and characteristic membranes (chorion, amnion, allantois, yolk sac)—it is highly improbable that this evolutionary advance can have happened more than once. And when, as recently, it is advocated, on the strength of a few osteological characters, that amniote origins occurred in half a dozen or so separate instances, we are beyond the bounds of credibility.

Nomenclature.—The Linnean system of nomenclature has proved a great boon to zoologists. This system of classification, although invented merely as a means of pigeonholing the various types of animals and plants, can be used in expressing relationships.

A few nomenclatorial rules are an obvious necessity. If, for example, a given name has been invented for one species within a genus, there should not be a second species with the same name or confusion will result. Further, if a name has once been coined and applied to a genus, that name should not be used a second time for a different animal; this is the creation of a homonym, and the second form should be given a new and different name. If a genus or species has been satisfactorily named, no second name should be applied to it. In case of dispute, it is reasonable that the first name used should be generally accepted and the second abandoned—the principle of priority. It was long agreed, however, that priority need not be adhered to if the first name given had been generally ignored and a subsequent name widely accepted.

During much of the last century zoologists got along, in general, happily as regards animal nomenclature, under, in essence, a series of "gentlemen's agreements" embodying a few general principles such as those just cited. Today, however, no such situation exists. Zoologists and paleontologists not particularly interested in minutiae of nomenclature are frequently puzzled and irritated by changes in the names—particularly generic names—of the animals in which they are interested. To what are such changes due? Why has confusion, and in some cases chaos, come about, in place of long-established stability? Curiously, this unsettled situation is in great measure the result of actions laudably designed to achieve the opposite result—stability!

At an international congress of zoologists in 1897 it was decided that in place of previous "gentlemen's agreements," there should be drawn up a code of rules regarding animal names, and that an International Commission on Zoological Nomenclature should be established to regulate nomenclatorial practices. In the code, the principle of priority was emphasized, although the commission was given power to make exceptions to priority if adherence to this principle would be disturbing.

For some time the earlier, essentially stable, situation in nomenclature continued. Later, however, particularly in the period following the First World War, there came a generation of students whose interests appeared to lie not in zoology itself—the study of animals—but in the curious field of the study of the *names* of animals. For such people a main interest in life became, it would seem, the routing out of obscure names having technical priority over those long in general use. The result of this sort of work tended strongly to produce exactly the opposite effect to that which the code aimed at—not stability, but confusion and change. If those who avidly sought for the dominance of priority had been concerned only with obscure forms of little scientific or economic importance, not much harm might have been done. Unhappily, this search for priorities tended to confuse the situation with regard to many familiar and important animals.

Let us take, for example, the living members of the order Primates, of which many are of very general interest to students of morphology, physiology, medicine, and behavior, as well as to the taxonomist and mammalogist. For such a group, most especially, stability is highly desirable. Unfortunately, strict application of the code and the actions of the commission have led to a great degree of instability and confusion.

In 1891 Flower and Lydekker's authoritative work, *Mammals Living and Extinct* (432), was published. In it the living primates were arrayed in 37 genera, bearing names which had long been in almost universal use and were then familiar to workers in a variety of biological fields. This list was an expression of an essential stability. A later standard work was Weber's *Die Säugetiere*, first published in 1904. He listed 48 genera of living primates, the increase in numbers being due mainly to subdivision of older, comprehensive genera. With two exceptions (among South American monkeys) all of the Flower and Lydekker names were used; the stable condition continued.

But by the time Weber's second edition (430) was published in 1928, the priority-seekers were at work, and Weber (although protesting one or two of the greater absurdities), dutifully changed the names of eight genera. The priority chase continued; this is reflected in Simpson's work of 1945 (431), less than two decades later, by which time (not counting five minor changes in spelling), six further changes had been made. We thus have a situation in which, within a few decades, a third of all the names long in use for primate genera have been discarded by the specialists and replaced by unfamiliar terms. The lower primates have been left generally unharmed; the carnage has, unfortunately, been greatest among the higher primates, monkeys and apes. Nearly half of the generally familiar names of such forms listed by Flower and Lydekker have been done away with; the chimpanzee, orang, langur, and marmosets are among those for which the technical experts insist that well-known appellations must be dropped.

What advantage has been gained by these changes? Nothing but confusion; without consulting an "expert," a younger worker is cut off, by the unnecessary changes in their names, from using the classic literature on many forms.

The most tragic act of the International Commission was that of Opinion 90. During the 1920's many workers on mammals became alarmed at the termite-like work of the priority-seekers, and proposed to the commission that the familiar names of some 16 "endangered" mammal genera be conserved, the list including the generic terms for such animals as the chimpanzee, orang-utan, common marmoset, peccary, flying lemur, the

Australian spiny anteater, and the manatee. The commission was then dominated by priority enthusiasts, such as Stiles and Stejneger. Five of the 16 names for which a plea was made (none of those listed above) were granted stability; suspension of the rules was refused for the others, which were, it was decreed, to take on names unfamiliar to zoologists in general.

Typical is the case of the walrus and the manatee. The former has been generally called *Trichechus,* although the alternative name *Odobaenus* was sometimes used; the manatee was almost universally termed *Manatus.* However, the priority-seekers discovered, to their delight, that although Linnaeus himself in 1766 and later editions used *Trichechus* for the walrus, his earlier edition of 1758 makes it appear that it then applied more properly to the manatee. What to do? A reasonable course of action might have been to "outlaw" *Trichechus,* since acceptable and unequivocal terms, *Odobaenus* and *Manatus,* were available for walrus and manatee, respectively. But no; apparently delighting in confusion, the commission decreed that *Manatus* should be sunk, and the manatee called *Trichechus.* This is the height of absurdity; no right-minded zoologist should or, I think will, submit to such a grotesque change.

"Titanotheres" and "oreodons" are among the commonest and best known of fossil mammals; even in geologic literature, the division of the fossil beds of the White River bad lands are divided into the "*Titanotherium* beds" and "*Oreodon* beds." But, by the laws of priority, neither of these two names are allowed in nomenclature. A picture of the first scientifically known specimen of a titanothere was published by a Dr. Prout of St. Louis, who did not, however, give it a generic name. A French paleontologist, Pomel, who saw the illustration (but as far as I know never saw a specimen of the animal) coined the term *Menodus*; Leidy, working with the actual specimen, called it *Titanotherium* and did not even know of Pomel's obscure and unjustifiable action until a decade later. Leidy described the first specimens of oreodonts from the White River; two names, *Merycoidodon* and *Oreodon,* were given, successively, to two early specimens; later, realizing that the two were generically identical, he chose to retain *Oreodon.*

In Leidy's day there were no "rules" regarding priority, and his opinion, as the major worker on White River fossils, that *Titanotherium* and *Oreodon* were the better names for the two forms concerned was accepted as reasonable by workers for many decades. Later, when the code was established, and a law of priority was included, it was made retroactive! In government affairs, in both civil and criminal codes, ex post facto laws are looked upon as abhorrent (in this country they are prohibited by the Constitution). Why should such retroactive decrees in science supersede the considered opinions of older workers? Should we adhere to doctrines of this sort? The classic answer to this is that, although in individual cases crimes have been committed in the name of priority, we should overlook them from the point of view that strict adherence to the "code" is more productive of stability and less productive of confusion than if individuals were to depart from priority in individual cases.

I very much doubt this. The continual pressure for adherence to priority has, I think, produced increasing confusion. I have been told by a friend (generally sensible) that "we get used to these changes." But who are "we"? Answer: the specialist. But animal names are not intended to be the private playthings of specialists; rather, they are for the use of biologists generally, who do not "get used" to continual change, but are confused—and rightfully irritated—by this nonsense.

Can this process of increasing confusion be stopped? Yes, it is possible as regards further change. Although the present commission on nomenclature contains a percentage of priority-seekers, the group as a whole is, today, a reasonable one, which is willing to consider a plea for the retention of a familiar name, even if a smell of priority impurity attaches to it. But bringing a petition to the commission for its consideration entails the presentation of documents for the preparation of which considerable amounts of work are involved. Few zoologists or paleontologists wish to spend their time in such quasi-legal nonsense, when we might better be employed in positive, constructive work in advancing knowledge of fossil animals.

It is thus possible to attempt, at least, to prevent future outrage, if one wishes to make the effort. But what of the damage already done by the commission? That, unfortunately, can never be completely remedied. For example, the orang, *Simia* of all older authors, has by now been called, even if absurdly, "*Pongo*" by a generation of younger workers, and even if the reasonable course of return to *Simia* were made, the duality of the name of the poor creature has already been embedded in the literature and can never be eradicated.

So far I have primarily discussed problems relating to generic names. But in the last revision of the "code," the legalists gave themselves powers over family names. This vastly increases the area for future turmoil and possible displacement of many a familiar name. I am told that it is probable that if the rules should be strictly enforced, half or more of the current names of mammal families might be displaced, and it is not at all improbable that the same would be true of names in other vertebrate classes.

A major question is this: are working zoologists necessarily bound by the rules of a group of specialists of narrow interests? I do not see that that is the case. The commission is in theory the servant of zoological society, not a set of masters whose orders we are bound to obey. Why should zoologists abide by decisions which are obviously nonsensical? The reply usually given is, as I have said, that not to do so would create chaos. But to my mind it is the nomenclatorial specialists who have been creating chaos. And it is my belief that a deliberate refusal to submit to the more absurd rules and "opinions" will make for a greater stability of zoological nomenclature than a blind (and generally grudging) adherence to absurdities of the past and (I fear) of the future.

Vertebrate structure.—A basic knowledge of vertebrate anatomy is a prime need for every worker; this to include not only the skeleton but "soft anatomy" as well. Lack of such knowledge almost inevitably leads to superficial treatment, and sometimes to laughable errors. To cite a single example: Lambe (321) in his otherwise excellent study of a large carnivorous dinosaur, *Gorgosaurus*, noted that in his specimen the teeth showed no signs of hard wear. He therefore concluded that *Gorgosaurus* must have fed upon soft material, and was a scavenger, rather than an active carnivore. Had Lambe, however, been a competent anatomist, he would have known that reptiles, unlike mammals, constantly replace their teeth, and the lack of wear did not necessarily mean more than that the individual had not had much to eat since his teeth were replaced; in fact, not improbably it may have died of hunger!

Bone.—Considerable work has been done in recent years on the histology of bone and other hard parts. Enlow and Brown (26), particularly, have been interested in bone structure in fossil and living tetrapods, but fish bones and scales have received a major share of attention. Gross, Denison, Bystrow, and others have been interested in the histology of dermal bones and scales. Ørvig, notably, has undertaken a series of studies on early types of bone and bonelike materials (39, 86); of great theoretical interest is the "lepidomorial theory" of scale growth and evolution proposed by Ørvig and Stensiö (63*a*; Stensiö 1962); a summary of much of the work done by the Stockholm school is given by Jarvik (8). Moss (1964) has published an excellent general discussion of skeletal history.

Before Stensiö's descriptions of cephalaspid anatomy (62, 63), it was generally assumed that the skeleton of primitive vertebrates consisted exclusively of cartilage and that bone entered the main line of vertebrate evolution only when the stage of the higher bony fishes, the Osteichthyes, had been reached. Ostracoderms and placoderms of a number of types were, to be sure, already described at that time; but they were inadequately known, little thought of, and presumed to be unimportant and aberrant forms (some, at least, were thought to be side branches from the higher bony fishes). Seeming confirmation of the presumed phylogenetic story was the embryological development of the skeleton in higher vertebrates. Dermal bones are generally formed directly from connective tissue, but elements of the internal skeleton are invariably first formed in cartilage, which is later replaced, in complex fashion, by bone. The ontogenetic replacement of cartilage by bone was looked upon as recapitulation of the phylogenetic story.

As Stensiö pointed out, his work put the whole story in a very different light. Here are fishes of ancient and lowly types, far below the level not only of the higher bony fishes but of gnathostomes *in toto*, and are yet possessed of good bony tissue, not only in the form of dermal bone but also, in some cases, of perichondral bone associated with internal skeletal structure. Further, as could be demonstrated, we find in many cases in lower vertebrate classes that, as time went on, the later members of a group showed not increasing ossification, but a decrease in bone and a strong trend toward a cartilaginous condition. This situation can be followed to a variable degree not only among ostracoderms and placoderms, but also among bony fishes, as in the case of the lungfish and chondrosteans; still further, as Watson (180), particularly, has shown, the main line of labyrinthodont evolution shows such a trend.

This general trend leads logically to the conclusion that the living cartilaginous fishes—cyclostomes and Chondrichthyes—are not primitive as regards skeletal structures, but are degenerate descendants of once bony ancestors. The ostracoderm story strongly suggests a descent of cyclostomes from bony ostracoderms; the pedigree of the Chondrichthyes is not so clear, but it is not unreasonable to believe, as does Stensiö, that the sharks and chimaeras are similarly descended from bony ancestral placoderms. (Note that there is little but cartilage in the internal skeleton of sturgeons and paddle fishes, and did we not have fossil evidence to prove the case, we would never suspect that they have been derived from highly ossified palaeoniscoids.)

Stensiö's conclusions seemed reasonable to me, and I have on various occasions (as in 1963 and 1964) elaborated on this theme of the antiquity of bone. These conclusions have not, however, met with universal acceptance, and with many there has been an inertia, an attempt to cling as long as possible to the classical theory of the early universality of the cartilaginous skeleton. Favoring such attempts is the inadequacy of our pre-Devonian fossil record. Many groups, for example, appear in early Devonian times without known fossil predecessors, and it is suggested that such forms were already in existence as cartilaginous types and first gained bony structures, one group independently of another, at this time. This is, of course, not impossible, and it might be, since the storage of calcium in the skeleton is physiologically important, that some environmental or internal physiological situation tended toward the development of calcium deposition in skeletal form independently in a number of groups. But the bony skeletons of early fishes are not merely calcium deposits (if this physiological need were the major factor, one would think a "calcium gland" ventrally situated in the abdomen would be the simple solution). Instead, we find that these early forms have highly developed bony skeletons of complex and varied natures. This complexity and variety implies a long antecedent history in skeletal development, rather than an immediate jump from no bony skeleton at all. The absence of antecedent stages is due, I feel certain, to the great dearth of continental deposits in the Lower Paleozoic—deposits representing inland waters in which I believe the ancestors of the Devonian fishes were evolving.

The earliest history of bone is, thus, almost unknown, although the presence of heterostracan material in the Middle Ordovician shows that bone, or bonelike tissues, is exceedingly ancient. Considering the trend toward bone reduction seen in many cases in the Devonian and later periods, one is tempted to extrapolate backward and assume that the ancestral vertebrates were highly ossified throughout the skeleton. Such a position, however, is quite surely unwarranted; after all, there had to be, in early Paleozoic times, a period in which bone was being acquired, before the process was reversed and loss set in. It is pure speculation, but one is tempted to suggest that bone first appeared in the form of membrane bone, as dermal armor protective in nature (against, I believe, eurypterid attack [Romer 1933]), followed by extension of bone deposition in essentially membrane form surrounding internal skeletal cartilages or their contained cavities (as is well seen in the cephalaspid head). From this point there might have been either further development with the initiation of true endochondral bone or the beginning of a reduction process.

The possible eurypterid-vertebrate relationship mentioned above was strongly impressed on me during a study of the occurrences of Silurian and Devonian fishes. That my idea was probably correct is further suggested by studies of Kjellesvig-Waering (1961) on eurypterids. He says:

> The numerous and persistent instances of the intimate association of pterygotids with primitive fishes throughout Silurian and Devonian time, as well as the presence of the greatly abraded teeth in the pterygotids herein described, point to the tentative conclusion that the fishes were a source of food to the great pterygotids.

If the embryological story of the replacement of cartilage by bone in the internal skeleton is not a recapitulation, why do we find it? The answer, although not to my knowledge discussed in earlier days, seems obvious (Romer 1942). Cartilage in vertebrates is to be looked upon as essentially an adaptation. Cartilage is found to some degree among invertebrates, and was presumably present in the ancestral vertebrates and in their chordate ancestors; but if it had not been present, it would have been necessary for the vertebrates to invent it to allow for the embryological development of the internal skeleton. The basic pattern of the skeleton is laid down in every vertebrate at a very early stage of the embryo. Bones cannot expand, and can grow in size only by the deposition of new tissue around the periphery of the bone already formed; growth of an osseous structure having complex relationships with other skeletal elements or "soft parts" is almost impossible without almost continuous disruption of organization. Cartilage, on the other hand, can grow by expansion and is a perfect material for laying down a tiny model of a structure which with growth can eventually be transformed into bone. If, in the evolutionary history of a group, the need for structural support or for calcium deposition furnished by bony tissue ceases to be necessary, it is possible to see how genetic change might eliminate the terminal bony stage of skeletal structures.

Not often taken into account, although of great theoretical importance in discussions of cranial structures, is the fact that much of the skeletal materials here are not formed from mesoderm, as are most bone and cartilage, but from "mesectoderm." As first observed by Platt two-thirds of a century ago, and most recently reviewed by Hörstadius (1951), cells from the ectodermal neural crest migrate downward in the head region to form the anterior portion of the braincase and the entire "visceral skeleton" of jaw cartilages and branchial arches. This unusual type of "mesoderm" formation is obviously important in the early evolution of vertebrates and, despite the fact that the embryology must forever remain unknown, should be kept strongly in mind in the study of early and primitive forms.

Dermal bone patterns.—The patterns of the dermal bones of the skull roof have played an important part in the study of lower vertebrate groups, since they are often the only diagnostic areas well preserved in a specimen and since the pattern is often dependable for diagnostic purposes. A number of patterns of quite different sorts are present in various groups of early fishes. Various attempts have been made to homologize bones between one group and another—between crossopterygians and actinopterygians, for example, or even between arthrodires and osteichthyans—generally without much success. See for example, Romer (1936), Westoll (165), Parrington (1956), and White (168, 1966).

One suggestion as to the remote ancestry of known diverse cranial patterns is that the ancestral forms had an armor formed of large plates, which became broken down in variable fashion in descendent groups. An alternative, to which I strongly tend, is that early fishes had a head shield composed of a mosaic of small bones; variable reduction in numbers and enlargement of the survivors in different groups led to eventual stabilization of patterns between which (as regards major groups) no homologization is possible. Stages in such a reduction process can be seen in the lungfish series and, as well, in rhipidistian crossopterygians, where the snout region may still show a variable mosaic but where the rest of the skull roof is nearly completely stabilized.

Vertebral structure.—Interpretations of vertebral structure were long dominated by the ideas of Gadow (1933), who postulated a basic pattern of four pairs of "arcualia." Attempts were frequently made, particularly in the case of tetrapods, to interpret the vertebrae of fossil forms in terms of Gadow's arcualia. I had long been unhappy over this situation, for no tetrapod shows evidence of more than three types of structure—intercentrum, pleurocentrum, neural arch. My doubts were shared by my colleague, Dr. Ernest Williams, who, in an able review (175), demonstrated that as regards tetrapods there is no evidence to support Gadow's ideas. As regards fishes, the situation is still obscure, owing in great measure to the fact that, teleosts apart, ossification of the vertebral centrum seldom occurs. Some light on the situation in actinopterygians may be shed by Schaeffer's current study of the development of the column in *Amia.*

For most labyrinthodonts and amniotes the history of the

"central" elements is clear. Inherited from crossopterygians, the primitive labyrinthodont centrum included two elements, hypocentrum or intercentrum, and pleurocentrum. In advanced stereospondylous labyrinthodonts the pleurocentrum is reduced and vanishes, so that the entire centrum is formed by the hypocentral ossification. In the line leading to amniotes, on the other hand, the hypocentrum is reduced to a small intercentrum and may eventually be completely lost, while the pleurocentrum enlarges to become the true centrum of amniotes.

But what of other amphibians? For the living orders the fact that the developmental story shows that the intervertebral disc lies intersegmentally, as does the intercentrum in amniotes, led Williams to conclude that here, as in amniotes, the body of the centrum is the pleurocentrum. Since the structure of the vertebrae of lepospondylous amphibians of the Paleozoic is similar to that of the living orders, Williams concluded that in these forms, as well, the centrum is a pleurocentrum.

I feel none too happy about this conclusion. Certain Paleozoic lepospondyls appear early in the Mississippian, with no trace of more than a single central element. But the pleurocentrum of typical rhipidistians and the oldest amphibians, the ichthyostegids, is a very small structure, and I find it difficult to believe that the pleurocentrum grew so rapidly as to completely supplant and eliminate the hypocentrum in (geologically) a short span of time. In the absence of embryological data one cannot be sure that we are not dealing here with a centrum formed by the hypocentrum, with the elimination of the small pleurocentrum (paralleling the stereospondyls). Or the ancient lepospondyls may derive from some group of fishes other than typical rhipidistians.

Tooth replacement.—Tooth replacement is of importance in the study of nearly all groups of reptiles, lower tetrapods, and higher fishes; we may note here the important work of Edmund (25). In a great variety of lower vertebrates tooth replacement (in contrast to mammals) continues throughout life. Seldom do we see in such forms (unless arbitrarily restored in a figure) an even tooth row; instead, we generally see an uneven arrangement, with some teeth large and fully developed, others small and obviously recently erupted, and still other tooth positions represented by toothless sockets in which replacement is taking place. In some instances the impression one gains is that of random irregularity, but in many cases close observation shows that the teeth in any jaw ramus appear to be divisible into "even" and "odd" series, each tooth in each series frequently showing a condition at any given jaw position contrasting with its neighbors. Further, in many cases, study of the two series indicates that waves of replacement were taking place in seemingly independent fashion in the two series, such waves usually appearing to move from back to front, but occasionally in the opposite direction. This curious phenomenon was long a source of puzzlement to many workers (including myself). Studies by Edmund—first on fossil material, and now confirmed by sequences of x-rays of living lizards—show that the solution is an unexpected but actually a simple one. Series of stimuli toward tooth formation pass along each row of tooth germs from front to back. If the spacing between traveling stimuli were exactly that of two tooth germs, there would be a perfect alternation of conditions between odd and even members of the series; but if the spacing were greater or less than this, the phenomenon of seeming waves of replacement would appear.

Brain.—We cannot, of course, expect actual brain tissue to be preserved in fossil form, but studies of cranial cavities (and endocranial molds) can tell us much concerning the nature of the brain and cranial nerves. Edinger was a leading student of paleoneurology; in 1929 (23) she published a summary of all then known in this field, and in 1937 a supplement; an annotated bibliography, bringing the work up to date was far advanced at the time of her death in June, 1967, and will, I hope, be brought to completion. The field has been an active one in recent decades; apart from numerous descriptions of individual finds, we may note a recent work on fossil mammalian brains by Dechaseaux (24) and one by Stensiö (1963) on lower vertebrates, abstracting his many detailed studies on cranial structures and contained brains and nerves.

Vertebrate ancestry.—It is reasonable to assume that the ancestors of the vertebrates among lower chordates were in general devoid of hard skeletal structures, and hence that our chances of encountering ancestral forms in the fossil record are very poor indeed. We should, however, have in mind concepts of the nature of the probable vertebrate pedigree to evaluate properly the position of the more primitive true vertebrates.

At one time or another almost every major invertebrate group has been suggested as one related to vertebrate ancestry. My own ideas on the subject, which I have discussed briefly at various times (Romer 1955*b*; Romer 20, pp. 27–30), are in general agreement with our present knowledge of lower chordates and of the structure and embryology of vertebrates. The theory starts, basically, with the fact that the structures of a vertebrate can, for the most part, be divided on embryological grounds into "visceral" and "somatic" components. The first includes, apart from certain structures of the head and pharyngeal regions, little but the digestive system and associated structures; the "somatic" animal includes the great bulk of the body—most of the skeletal and muscular systems, the central nervous system, and sense organs.

This suggests the possibility that the far-distant metazoan ancestor of the lower chordates and vertebrates was not, as one might think, an active form, but a simple sedentary creature composed of little but a digestive tract and an apparatus for gathering food particles from the surrounding water—that is, a simple "visceral" animal. Such forms are present today as the bryozoans, brachiopods, and, among the echinoderms (with whom there are other suggestions of chordate affinities), the crinoids. These are simple stalked little animals with ciliated arms along which captured food particles are brought to the mouth. Exactly the same sort of structure is found in the pterobranchs, lowest of chordate types.

A stage upward from this would be the development of a system of gill-slit filter-feeding. This method is characteristic of all higher chordates (and even in the lamprey larva among true vertebrates), and the gill system is the most basic characteristic of the chordate phylum. A major advance toward vertebrate conditions was the initiation of motility by the development of a muscular swimming tail and sense organs and central nervous system for the regulation of navigation. Garstang's (1929) suggestion that, as seen in some ascidians, this first evolved as a larval tadpole structure which was later retained into the adult stage has become an increasingly acceptable explanation. With this addition, little more was needed beyond the acquisition of a skeleton to transform the chordate into a vertebrate; and some of the older ostracoderms, particularly the cephalaspids, seem still close, apart from the presence of a skeleton, to what we would expect of an advanced pre-vertebrate.

Remains that can be confidently attributed to chordate or hemichordate ancestors are rare in the fossil record. Kozlowski and others have claimed that the Paleozoic tubelike structures termed graptolites are the tubes of pterobranch-like forms, and certain of these ancient forms, it is suggested, may have been actual pterobranchs. There are no sure traces of fossils representing the other known lower chordate types—acorn worms, tunicates, *Amphioxus*—none of which has fossilizable skeletal structures today. A curious fossil from the Silurian of Scotland, *Ainiktozoon* (Scourfield 1937; new material is being described by Ritchie), which looks like nothing ever dreamed of in heaven or on earth, has been thought to be some type of lower chordate. It may or may not be. There is no real evidence.

Conodonts.—In recent decades invertebrate paleontologists and stratigraphers have been much interested in conodonts—tiny tooth- or jawlike structures, mainly Paleozoic in age—because they have proved useful in correlation; Lindström (1964) has recently published a summary of knowledge of them. Little attention has been given to their zoological position, but since they do not fit well into any living invertebrate phylum, those working with them—mainly invertebrate paleontologists—have tended to claim that they are vertebrates. But neither do they fit into any reasonable scheme of vertebrate history nor into any pattern of vertebrate structure, and vertebrate workers have pushed them back with equal vigor. At one time or another assignment to almost every main invertebrate group has been suggested (Rhodes 1954). Recently Fahlbusch (1964) has maintained that they are not animal remains at all, but algal growths! I find this hard to believe, but not much harder than many other theories advocated. Gross (1954, 1957) has suggested that although definitely not vertebrates, they might perhaps represent some extinct group of lower chordates—a solution that might appease invertebrate workers and yet cause no alarm to vertebrate paleontologists. But I think it still more probable that they represent some ancient invertebrate stock now entirely extinct. Even today there are various metazoan invertebrate types completely or nearly completely soft-bodied, whose existence would never have been suspected had they failed to reach the Recent. Perhaps some future discovery of a Paleozoic deposit of the Burgess-Shale type may give us knowledge of the true nature of conodonts—or perhaps not.

Archaeognathus consists of a single specimen from the Ordovician claimed by invertebrate paleontologists (Miller, Cullison, and Youngquist 1947) to be remains of a fish. Whether or not this is a conodont (it looks like one to me), there is no adequate reason for considering it a vertebrate.

Early vertebrate environment.—The question of whether early fish evolution took place in fresh or salt water has been, and will long continue to be, a matter for debate. Presumably the sea was the original home of animal life, and it was hence long assumed that the sea was also the early home of the vertebrates. Considerable evidence, however, exists which throws doubt upon this assumption. Homer Smith, in various works, maintained that kidney structure and function indicate that the early vertebrates were fresh water in origin. Almost every vertebrate maintains in solution in its body fluids a series of salts comparable in general nature to those present in sea water, but in lower concentration. The most generalized type of vertebrate kidney structure is one which is functionally adapted to resist dilution of body fluids in a freshwater medium; marine fishes, which must resist dehydration and excess of salts due to osmosis, have adaptations to this end; but such adaptations are varied in nature and strongly suggest that a number of fish types have independently invaded the seas. Homer Smith's conclusions have been generally accepted, but a minority, notably Robertson, have opposed his ideas, mainly on the basis that one vertebrate, *Myxine*, shows no regulation of salt content of body fluids.

On the paleontological side, Grove and I in 1935 reviewed the American evidence concerning the older fossil fish finds and concluded from this that the earliest vertebrates were dwellers in fresh water, and that only during the Devonian was there any major trend toward the development of marine fish faunas. This conclusion did not long remain undisputed, however, and Gross (1950), White (1958*b*), Robertson (1957), and Denison (38) have all argued in favor of salt water; it is difficult for me to attempt objectivity, but despite the 4–to–1 poll against me, I still feel sincerely that, when considered broadly, the freshwater viewpoint is that which best explains the known data, and I hope on some future occasion to again discuss this interesting question. (Some notes are given in chapter 28.) Here, I will merely note a few facts worth careful consideration.

1. Repeated peneplanations have resulted in almost complete elimination from the known geologic record of Lower Paleozoic continental deposits. As far as I am aware, there are none in the Cambrian, almost none in the Ordovician, and very few in the Silurian until late in the period—as the Devonian is approached and reached. In consequence, any vertebrates which might have existed in fresh waters in the Lower Paleozoic can in general be found only in surviving deltaic or estuarine deposits or in coastal waters into which their remains might have been carried.

2. In the Silurian, vertebrates are rarely found in typical marine deposits; their remains increase as we approach continental areas in time or space.

3. A great variety of gnathostomes, already highly developed and differentiated, appear early in the freshwater Devonian. Except for fragmentary remains of acanthodians, there is absolutely no trace of them in the dominantly marine Silurian.

2

Primitive Jawless Vertebrates

Modern study of primitive vertebrates began, essentially, with Stensiö's classic monographs on cephalaspids in 1927 and 1932 (62, 63). Since that time, work on the various ostracoderm groups has been unabated. A considerable number of recent works dealing with one or another specific type or group are noted below; more general treatments include Stensiö's summaries of ostracoderms for the Grassé Traité de Zoologie in 1958 and the Piveteau Traité in 1964 (63*a*) and a comparable summary by Obruchev (1964) in the Russian Treatise. Westoll (36, 67) discussed the evolution of ostracoderms, as did Obruchev in 1945 and Lehman in 1959 (7); Watson (69) has commented on important points in ostracoderm structure and history. Important with regard to the consideration of ostracoderms and of other early vertebrates as well are studies on bone and scale structure, such as the lepidomorial theory of the Stockholm group (Stensiö, 1962, etc.), studies of bone structure by Ørvig (1965*a*, etc.) and Gross (1961, 1966, etc.), and Stensiö's (1963) reconstructions of the brain and cranial nerves in lower vertebrates.

Osteostraci.—A number of works about the cephalaspid and related types have appeared in the postwar years, such as a lengthy study by Wängsjö (65), several papers by Denison (66, 68, 1951*a*), including discussion of their classification and evolution, a consideration of cephalaspid structure by Watson (69), description of British material by White (1950), and a new classification by Stensiö (63*a*). Stensiö's classification is complex, involving a division into two orders and a considerable number of families. The relative uniformity of the cranial pattern among osteostracans suggests that a more conservative classification is preferable, and I have used one based on those of Wängsjö and Denison. Stensiö's classification relies primarily on internal characters; hence, lamentably, it is generally impossible, in his scheme, to identify the genus (or sometimes even the family) of a form without investigation of internal anatomy. Even the "type" genus *Cephalaspis* is left *incertae sedis*; White (1958*a*), however, has been able to pin this type down. Certain characters upon which Stensiö relies strongly, such as the size of the mouth and the consequent crowding (or lack of crowding) of the gill slits, do not seem to be of major importance in themselves.

Except for occasional details, Stensiö's original description of the cranial structures of cephalaspids has proved to have general validity, as has his interpretation of much of the soft structural features contained within the bony framework of the cephalaspid "head." Certain of his interpretations, summarized by Watson (69), are open to question and reinterpretation. One striking feature of the cephalaspids not represented in the modern lampreys is the tesselated plate areas present on either margin and on the posterior central region of the cranial shield, to which large tubes extend outward from the posterior portion of the cranial cavity. Stensiö at first hesitated between interpretation of these structures as sensory or as electric organs, but finally decided for the latter. It becomes, however, increasingly probable that the sensory interpretation is the more likely, as both Westoll (67) and Watson (69) conclude. For the development of an electric "shocking" structure, a considerable volume of space is necessary; beneath the lateral plate areas such space is lacking. The presence here of sensory structures seems much more probable; these presumably had a function analogous to lateral line organs, the tubes to the endocranial cavity being filled with liquids comparable to perilymph, transmitting water vibrations to the internal ear.

It is obvious that the well-defined gill areas of the cephalaspid "head" are comparable not to the gills of higher fishes, but to the gill structures of lampreys. Here, however, the problem is whether these structures are of the nature of the adult lamprey gill pouch or the gill structure of the ammocoete larva. Stensiö believes that the pattern was that of the larva, with water flow caused by ciliary action; Watson adduces arguments suggesting that the structure was that of the adult, with a pumping musculature involved. The number of gills, typically ten pairs, is greater than that found in higher fishes (but comparable to conditions in some cyclostomes). Obviously, the more posterior gills are posterior to the hindmost present in typical jawed fishes; and, on the other hand, it is certain that one or more of the most anterior gills are anterior to the first slit of higher fishes. The evidence from the embryology and anatomy of modern fishes suggests that in the development of jaws, at least one and possibly two anterior gill slits present in the pre-gnathostome stage had been eliminated. The first gill slit of modern jawed fishes is supplied by the "hyoid" or facial nerve (VII). The trigeminal proper (V_2, V_3), it is generally agreed, once served a gill slit lost with jaw enlargement, and it is possible that the profundus nerve (V_1), much reduced in gnathostomes, served a still more anterior gill slit, leaving, perhaps, the little terminalis nerve as the original nerve of the snout region.

The cephalaspid head tends to show that these deductions, based on modern gnathostomes, are well founded. Although it is sometimes said that, according to Stensiö, only one "lost" pre-spiracular pouch is present, Stensiö in fact says (62, p. 166) that it is the third cephalaspid pouch which is hyoidean—i.e., supplied by the facial nerve, leaving two more anterior ones to be lost in the assumption of the gnathostome state.

Typical Devonian cephalaspids have a pair of "horns" extending out and back at the posterior corners of the head shield, within which are sheltered scale-covered "paddles" in the proper position for pectoral fins. Certain forms, such as *Tremataspis*, lack both "horns" and "paddles," and the solid "head" structure extends further back along the body than the presumed pectoral region. Adherence to an extreme form of the fin-fold theory would demand that all pectoral fins have evolved from structures present in a common ancestor, and that, hence, the cephalaspid "paddles," if they are true pectoral fins, were present in the ancestral osteostracan and that their absence in such forms as *Tremataspis* is secondary. Stensiö here (as in the case of the arthrodires) tends to consider the presence of "fins" as primitive, lack of them (with a consolidation of the anterior trunk segments with the "head") as specialized. This is the reverse of the sequence suggested by the stratigraphic distribution—which Stensiö flatly rejects. The paleontological evidence suggests that "horns" and "paddles" evolved within the osteostracans (Heintz 64; Westoll 67), and that, hence, the "paddles" are not strictly homologous with pectoral fins of other fishes. They may still be thought of as homologous in a broad sense, however; Westoll has pointed out that in various ostracoderms there is a strong trend for development of paired ventrolateral balancing structures along the flanks, and that finny extensions may have developed from this source independently in a number of cases.

Although the body shape and proportions of cephalaspids are specialized rather than primitive, their proportions and presumed mode of life seem suggestive in relation to the story of chordate evolution which I incline to accept—namely that an early chordate was essentially little more than a food-filtering apparatus, to which eventually a motile tail structure was appended and a series of sense organs supplied, giving motility to the feeding apparatus first in the larval, and later in the adult stage. The cephalaspid, despite its various specializations and advances, gives very much the picture of this theoretical ancestral vertebrate—the "head" (apart from sense organs) being mainly a large food-filter to which was appended a muscular tail able to transport the filtering device to places where satisfactory nourishment could be obtained.

Anaspids.—These minnow-like freshwater ostracoderms of the late Silurian and Devonian were first described by Traquair from Scottish specimens; the structure of the head region was obscure, and since the tail was asymmetrical, it was presumed to be a typical heterocercal type, with the distal end of the body tilted upward. Although this pose placed a series of spines which appeared to be protective in nature along the bottom of the fish, instead of dorsally, where they should be, it was generally accepted by early workers, except Jaekel (a scientist noted for eccentric—but often correct—beliefs). In 1924, however, Kiaer (70), on the basis of new and better Norwegian materials, showed that Jaekel was right; the tail was a reversed heterocercal one, with the body axis tilting downward, not upward.

No internal ossifications are known in anaspids. From the

pattern of dermal bones and scales, however, Kiaer was able to make out much of the general structural pattern—including the presence of a single dorsal nasal opening—and reasonably deduce that the anaspids were cyclostome relatives.

A number of additions to our knowledge of the group have been made in recent decades. Parrington (72) has given a valuable general discussion of anaspid life and probable habits. Of interest was the description by White (1946*b*) of *Jamoytius* of the late Silurian, in which he believed bone to have been absent, but myotome impressions present. Although too late to be truly ancestral, this form, White suggested, was a morphologically ancestral one, bone having only evolved at a relatively late stage in the evolution of the group. It has, however, been shown since (Ritchie 74) that these impressions are those of bony scales, rather than myotomes, and *Jamoytius* has thus lost much of its theoretical importance. Stensiö in 1939 (71) described from the early Upper Devonian of Scaumenac Bay a new genus, *Endeiolepis* (of which a further imperfect specimen is present in the Harvard collections), in which a fin-fold (of especial interest to adherents of the fin-fold theory of paired fin origins) was present along either flank. No such structure was thought to be present in other anaspids; recently, however, Ritchie (75) has demonstrated that such a fold was present in *Pharyngolepis* and probably in anaspids generally. Heintz (73), Smith (1957), and Stensiö (1958, 63*a*), as well as Ritchie, have made new restorations, particularly of the head region, and Stensiö has made interesting hypothetical restorations of the possible internal structures. Smith (1956) presented evidence suggesting preservation of some internal skeletal structures which Ritchie, however, believes illusory. A tuberculated ventral plate in the mouth region was suggested by Stensiö (1958) as an incipient "rasping tongue" device indicative of pre-lamprey conditions. Ritchie, however, believes that this plate was part of the mouth border.

Heterostraci.—In early days, most specimens pertaining to this group were assigned to the single genus *Pteraspis* (except for isolated ventral shields termed *Scaphaspis*). Most early finds were from Europe. But as with the Osteostraci, our knowledge of the distribution and variety of members of the group has increased enormously. Kiaer and Heintz (76), for example, described quantities of heterostracans from Spitzbergen; in North America they are abundant in the Rocky Mountain area (Denison 79), as well as in the Appalachian region; Denison (84) has recently described a fauna from Alaska; Thorsteinsson (1958) has discovered abundant materials, now being described, from Cornwallis Island in the far north, and Dineley (1966) reports further quantities of materials on nearby Somerset Island. Stensiö's (63*a*) recent summary for the Piveteau Traité shows something of the complexity of the present situation. Gross (1963) has excellently redescribed *Drepanaspis*; Tarlo (1961, 1962, etc.) has published extensively, particularly on Polish forms; White (1963, 1946*a*, etc.) has mainly studied forms from the Welsh border, notably *Pteraspis* (*sensu lato*); Denison (81) has reviewed the Cyathaspidae. Liu (1965) has given a preliminary description of a Chinese Devonian fish, *Polybranchiaspis*, which appears to be a good heterostracan in most regards, but in which the gills, in contrast with proper heterostracans, are said to open separately outward.

In the Heterostraci we are dealing with creatures of a type very different from the osteostracans and anaspids. Those two groups are certainly related, despite the differences in their proportions; but the Heterostraci are obviously quite distinct. As in the anaspids, internal bony structures are absent, but here not merely can deductions be made from the outer surface of the dermal elements, but, in addition, much can be deduced—regarding, particularly, the gill pouches and the brain cavity—from impressions on the inner surface of the dermal armor. In striking contrast to the osteostracans and anaspids, there are no individual gill openings; instead, the bony armor forms an operculum of a sort, and there is but a single gill opening on each side. Further, in contrast to the situation in the two groups previously considered, there is no single dorsal nostril opening.

Stensiö in his 1927 treatment of ostracoderms (62), in which he reasonably concluded that cephalaspids and anaspids were lamprey relatives, expressed the belief that the Heterostraci were relatives of the hagfishes, and has since maintained this position (63*a*). Like most other workers (e.g. Watson 69; Heintz 78; Denison 81), I cannot agree that this is the case.

For one thing, since the Heterostraci were even in the Paleozoic widely separated from cephalaspids and anaspids, it assumes that such cyclostome specializations as the peculiar tongue evolved twice quite independently—which I find difficult to believe. Jarvik (1965*a*) points out that differences in detail exist between the "tongues" of lampreys and hags, but the basic structure appears to be the same. Further, Stensiö's theory assumes that, like the hags (and lampreys), the Heterstraci were monorhinal. For this there is no proof. The single hagfish nostril is terminal. The Heterostraci have no terminal opening, and water to the nose must have entered through the mouth cavity. In his most recent treatment Stensiö has made interesting and elaborate reconstructions to show that possibly there might have been a single nasal pocket deep within the heterostracan head region. But there is no positive evidence to support this reconstruction and, on the other hand, as Heintz, particularly, points out, there are definite pockets at either side of the mouth in which it is much more reasonable to believe paired nostrils were lodged. It would seem that the heterostracans were in no way related to cyclostomes (apart from being agnathous), and were, to some slight degree at least, more closely related to the central stock leading to higher two-nostriled fishes.

The Heterostraci are still the only vertebrate group definitely known from the Ordovician (the toothlike structures from the Russian Ordovician first described by Rohon give little indication of their zoological position). The heterostracan remains from the Ordovician were originally described from the Harding Sandstone of Colorado, but various geological investigations and further exploration, by Ørvig (77) and by other geologists and oil workers, have shown them to be present in a series of formations of late Middle and early Upper Ordovician age extending northward from this point through eastern Wyoming and (subsurface) eastern Montana—positions, incidentally, lying adjacent to a presumed Ordovician land mass running north and southeast of this line. Ørvig has described a new genus, *Pycnaspis*, showing a fair expanse of armor although, unfortunately, not enough to give an idea of body shape.

Stensiö has, in 1958 and 1964 (63*a*) given an elaborate classification of heterostracans, in which he raises the group to superordinal level and ranks the forms considered in ten orders. Tarlo (82) gives a classification nearly as complex, with eight orders, the only differences being that the corvaspids and drepanaspids are united in a single order, and the Turiniida, presumably coelolepids (Westoll 36), are eliminated. I see no marked advantage in "blowing up" the subgroup differences to ordinal status, and although having for the most part followed the subdivisions of Stensiö and Tarlo, I have considered their "orders" in general to be of family status only.

Coelolepids.—Apart from reports and descriptions of scales, no new data of importance have been presented on coelolepids (thelodonts); we still do not know whether the Silurian and Devonian materials included here pertain in part or entirely to heterostracans, are young individuals of typical ostracoderms (Westoll 67), or in part at any rate, pertain to a group that might—or might not—be of a more generalized type than other known ostracoderms and perhaps closer to the ancestry of higher vertebrates.

3

Archaic Gnathostomes

Placoderms.—In retrospect, it appears to me that (oddly enough) our modern views of the nature and unity of that great group of armored Devonian gnathostomes, the Placodermi, were arrived at in great measure as a result of Watson's monograph of 1937 on acanthodians (123). This was far from his main objective in this paper which, apart from a thorough description of the acanthodians, proposed to show that they, the arthrodires, and a variety of other Paleozoic forms were a united group, which he would term the Aphetohyoidea, and that in these fishes a primitive condition existed in which the hyomandibular took no part in jaw suspension. His main theses have proved not to be true. Both in acanthodians and the arthrodires and their relatives, the hyomandibular may support the jaw joint; acanthodians show no indications of any close relationships to the arthrodires and their kin; the term "Aphetohyoidea," as both cumbersome and (as was suspected and presently realized) inappropriate, was soon generally replaced by Placodermi, a term proposed by McCoy a century ago.

But it was Watson who first made it clear that in the vast array of Devonian types which (omitting the acanthodians) could properly be included in the placoderms, a common and unique body plan was discernible, though highly varied, in forms as far apart as typical arthrodires, antiarchs, and the skatelike *Gemuendina.* The diagnostic character is an armor consisting of discrete "head" and trunk segments and a shoulder region from which projected (unless lost in late arthrodires) a shoulder spine, which is the best placoderm trademark.

Stensiö has been the outstanding major student, for the past forty years, of this great group of Devonian fishes. The first of his important studies on members of the group was his 1925 description of the skull of *Macropetalichthys* (101), to be followed in later decades by numerous other works (91, 95, 111, 1947 [*passim*], 1963 [*passim*]) and, most important, his recent elaborate monographs on the placoderm pectoral fin and shoulder girdle (97) and on the head, of which only the first part has as yet been published (95). It is expected that a summary of his current views will appear shortly in a section of the Piveteau Traité dealing with "lower" gnathostome fishes.

Stensiö tends to use the term Arthrodira for this entire assemblage, calling the arthrodires of common usage the Euarthrodira; I think it preferable, however, to use Placodermi for the group as a whole and restrict Arthrodira primarily to the assemblage familiarly known by that term, rather than stretching it to such far-off types as (for example) *Gemuendina* and the antiarchs. The Placodermi are a notably isolated group. Except that a few may have survived into the Carboniferous, they are strictly confined to the Devonian period, where they are exceedingly prominent in marine strata. Nothing is known of their ancestry; there is not the slightest indication of relationship to groups—acanthodians or ostracoderms—known from pre-Devonian deposits. Nor are they at all antecedent to advanced bony fish. They may well be ancestors of sharks and chimaeras; but even in the case of the chimaeras, the evidence is slender, and there is no known connecting link of any sort with the elasmobranchs.

Arthrodires.—Most abundant of placoderms are the true Arthrodira—the Euarthrodira of Stensiö's nomenclature. During the past two decades at least a score of important papers have been written on them, as well as numerous minor ones; apart from the major works of Stensiö, already mentioned, we may cite as examples papers by Bystrow (87), Denison (92), Gross (1961, 1962), Heintz (1962), Kulczycki (1957), Lehman (99, 1962), Miles and Westoll (1963), Obrucheva (1962), Stensiö (91, 1963), Westoll (88, 1958*a*), Westoll and Miles (1963), and White (93); Westoll and White, particularly, have discussed evolutionary problems. We now know well over 100 genera of these unusual fishes. They flourished throughout the Devonian, and such Upper Devonian deposits as those of the Cleveland Shale and Wildungen in Germany have a wealth of genera. We have no clue as to the cause of their seemingly sudden extinction, to be succeeded in the Carboniferous seas by the elasmobranchs.

Lower Devonian arthrodires, most abundantly present in Spitzbergen, are characterized in general by a long stretch of body armor, small fins emerging from a small opening in the armor of the shoulder region, and very long, curved pectoral spines. Although there is some overlap, Middle Devonian forms, such as *Coccosteus*, generally have a shorter trunk armor, a larger pectoral fin emerging from a larger opening, and a reduced pectoral spine; in the late Devonian there is further armor reduction, the pectoral fins (as far as known) are still more developed, and the spine reduced to a nubbin or lost entirely.

This geological sequence, first clearly established by Heintz in 1931 (89; cf. Heintz 1938), seemed to most workers an evolutionary sequence as well, and although some (as White 93) are inclined to think that the spine development in some Lower Devonian forms is excessive and off the main track, most divide the group into Arctolepida (or Dolichothoraci), including mainly Lower Devonian forms, presumably primitive, and Brachythoraci, for the seemingly more advanced arthrodires of the Middle and Upper Devonian. There is no sharp cleavage between the two groups, and it may well be that the two terms represent merely grades, with advance from arctolepid to brachythoracic levels occurring independently in a number of lines. Among the Brachythoraci such forms as *Coccosteus* are relatively conservative. Others, particularly in the late Devonian, such as "*Dinichthys*" (properly *Dunkleosteus*, *fide* Lehman 99) are highly specialized; a division is often made here between Coccosteomorphi and Pachyosteomorphi.

Stensiö (97), in his studies of the pectoral fin and shoulder girdle, comes to a conclusion about arthrodire evolution startlingly different from that just outlined. Despite the fact that the late Devonian forms with broad fins and spine reduction appear to be most closely allied to other and seemingly ancestral brachythoracians, he would sharply split off these forms (the Pachyosteomorphi), as Aspinothoracidi, from all the remaining placoderms—not merely from other brachythoracians and arctolepids, but from all the more curious placoderm assemblages as well, lumping all these as Spinothoracidi. Despite my admiration for Stensiö's splendid work, I (and I think most other students of early fossil fishes) must dissent strongly from this arrangement. It apparently is based on the assumption that a well-developed, broad-based fin of "fin-fold" type must be primitive. But acceptance of this thesis makes it necessary to assume that the geological record presents the arthrodires upside down. This is difficult to believe. On the basis of the abundant data available, it is more logical to accept the evidence at its apparent value and assume the gradual evolution of the arthrodire "main line" from early primitive arctolepids with narrow-based fins and well-developed spines, up through the more primitive brachythoracians to advanced "aspinothoracians" in the late Devonian.

Acceptance or rejection of Stensiö's proposal may be, of course, strongly influenced by one's theoretic beliefs of the origin and evolutionary history of paired fins. No one today would, I think, adhere to the classic Gegenbaur archipterygial theory of the origin of paired appendages, but the fin-fold theory in one form or another is still influential (cf. Jarvik 1965*b*). In its basic form it merely states that median and paired fins are basically similar in structure and hence in origin, and that the paired fins developed as lateral folds, much as dorsal fins developed medially, as stabilizing structures. In its extreme form the fin-fold theory assumed a continuous fin-fold, and from this there has been very generally retained the concept that a broad-based fin with a row of parallel supports is the gnathostome "archetype." This idea receives support from the fact that such fins are present in the oldest sharks, in reduced form in early actinopterygians, and in late Devonian arthrodires. But even apart from the seeming evidence of the history of arthrodires proper, a review of what is known in other placoderms tends to suggest that narrow-based fins are primitive for placoderms—the pectoral fin is narrow-based in ptyctodonts and in all rhenanids and petalichthyids in which this structure is preserved; and even in *Gemuendina*, despite the great expansion of the pectorals distally, there is obvious considerable proximal concentration.

Theories which postulate a broad-based fin as the archetypal form are based on the assumption that paired fins have evolved once, and once only. This is far from certain. In particular,

Westoll (67, 88, 1958*a*, etc.; cf. Nursall 1962) has pointed out that, looking at the question broadly, there was a general trend among early vertebrates towards the development of outgrowths along the flanks, and that it is highly probable such structures may have developed a number of times; surely the pectoral "paddles" of the Osteostraci and the lateral folds of the anaspids are developments parallel to but not directly connected with paired-fin development in gnathostomes. It is not at all impossible that fin development in placoderms took place independently of a comparable evolutionary development of fins in "higher" fishes.

Certain placoderm types, such as *Holonema* and its relatives, and *Heterosteus* and *Homosteus*, are somewhat divergent from the general stream of arthrodire evolution, but they are nevertheless unmistakably part of the arthrodires proper. Others are more remote. *Phyllolepis*, for example, is a depressed form with reduced armor which has the appearance of an ostracoderm and was long thought to be one.

Ptyctodonts.—Specialized in quite another direction, but clearly showing the basal arthrodire pattern (as Dollo first recognized) are the ptyctodonts. These forms are most commonly represented by their tooth plates, apparently nature's first experiment in the development of a vertebrate dentition adapted to treating "shellfish." Good specimens are rare; notable are descriptions of *Rhamphodopsis* by Watson (106) and *Ctenurella* by Ørvig (107). Various writers have suggested, on the basis of tooth-plate similarities, that ptyctodonts gave rise, with loss of bony armor and other changes, to the chimaeras of Mesozoic and later times. Ørvig (107) is the most recent strong advocate of this theory; Westoll (108) agrees but differs on various points of interpretation. There is, of course, a long gap between the last appearance of the ptyctodonts at the base of the Carboniferous and the appearance of the chimaeras in the Jurassic. It has been frequently suggested that the tooth plates of the Cochliodontidae of the late Mesozoic represent transitional forms between ptyctodonts and chimaeras. As noted later, Patterson utilized, but in another fashion, the cochliodont evidence, which indicates to him that although chimaeras may be derived from placoderms, the ptyctodonts are not ancestral.

Antiarchs.—The antiarchs, with such familiar forms as *Bothryolepis*, *Astrolepis*, and *Pterichthyodes* (*Pterichthys*), are a group immensely abundant in the later Devonian (nearly all in fresh waters); for example, at such a productive locality as Scaumenac Bay in Canada, probably two of every three fish specimens encountered are *Bothryolepis*. Stensiö (111) gave an exhaustive account of this genus in 1948, Karatajuté-Talimaa (1963) has given an extensive account of *Astrolepis* from the Russian platform, Liu and colleagues (1958, 1963) and Chang (1965) have recently described new forms from China, Gross (1965*a*) has recently suggested a revision of antiarch classification, and Watson (1961) has made some interesting suggestions regarding the group. We have no good explanation for the spectacular success of these forms for a period. One possible factor may have been the presence of lungs, which has been clearly demonstrated by Denison (110). However, most of the antiarchs' freshwater contemporaries were bony fishes that were also presumably lung-bearing.

The ancestry of the antiarchs is quite unknown. The "flippers" suggest the long pectoral spines of acanthaspids (Westoll 36, 88), but there are no transitional forms between fixed and free spines. The close-set eyes and dorsal nostrils suggest the rhenanids; but in the only adequately known rhenanid, *Gemuendina*, fin development has followed a very different course.

The "funny-fishes."—There remain for discussion an assortment of Devonian forms, mainly from the marine Lower Devonian, of curious pattern and without, at first sight, much evidence of relationship to other fish groups (for which reason some of us discussing them agreed at one time to provisionally call them merely the "funny-fishes"). Armor may be highly developed or greatly reduced; head patterns vary greatly; fins (where known) also vary from small, fan-shaped structures to enormous skatelike pectoral expansions. When their structure is analyzed, however, all are found to have, in some fashion or other, the characteristic placoderm features of dual head and trunk sets of armor, and all well-known forms possess the placoderm "trademark" of a pectoral spine—a point first emphasized by Watson (123).

Most of these forms are poorly known; it is to be hoped that better descriptions of many will be available presently with the publication of the second part of Stensiö's "Anatomical studies on the arthrodiran head." Stensiö (97) in 1959 erected five orders for the dozen or so genera concerned: Acanthothoraci for *Palaeacanthaspis* and *Dobrowlania*; Radotinida for *Radotina*; Rhenanida for *Gemuendina*, *Asterosteus*, and *Jagorina*; Petalichthyida for *Macropetalichthys*, *Lunaspis*, *Notopetalichthys*, etc.; Stensiöellida for *Stensioella*, *Nessariostoma*, and *Pseudopetalichthys*. Independently of this, I had concluded that on the basis of current evidence these "funny-fishes" appeared to be assignable to two groups; the Rhenanida, including *Palaeacanthaspis* and *Radotina* as well as the forms included in the rhenanids by Stensiö, and the Petalichthyida, to include (questionably) *Stensioella* and relatives with the petalichthyids proper. This simple arrangement, however, does not differ to any great degree from Stensiö's beliefs, since he states (97, p. 7) that "*Radotina* . . . is clearly related to the Dolichothoraci, Acanthothoraci and Rhenanida," and that his order Stensiöellida "is most certainly related to the Petalichthyida," hence giving the possibility of "lumping" his five orders into two comparable to mine. Obruchev (1964) has a somewhat different arrangement. Like myself, he separates the Rhenanida and Petalichthyida from the proper arthrodires, but includes the *Stensioella* group in the rhenanids, despite the fact that they do not have the closely appressed eyes characteristic of other rhenanids. To the petalichthyids he attaches the phyllolepids and ptyctodonts, although I can see little positive reason to bracket them here.

Of the Petalichthyida, the head shield of *Macropetalichthys* has been known for over a century, and Stensiö's first major study of Paleozoic fishes (101) was the description of the *Macropetalichthys* braincase some 40 years ago. For a long time we knew nothing of this widespread Middle Devonian form except the "head." Studies by Heintz and, most recently, by Gross (102) of *Lunaspis*, a Lower Devonian forerunner, revealed the general structure of a member of this group, however. We are dealing with forms showing the basic pattern of placoderm armor and, in *Lunaspis*, of a complete covering of the body and tail with bony scales, a condition (presumably primitive) preserved in few other known placoderms. The pattern of the petalichthyid armor is, however, rather different from that of the arthrodires and other placoderms mentioned earlier, and we are obviously dealing with a branch of the placoderm stock of primitive nature but early divergent from the line leading to arthrodires. Very probably related are *Stensioella* and *Paraplesiobatis,* described originally by Broili and recently restudied by Gross (103, 1965*b*); they are inadequately known, but appear to show some further degeneration of armor.

Of the rhenanids, *Gemuendina* of the Lower Devonian has been known since the days of Traquair, was further described by Broili, discussed to advantage by Watson (123) and most recently redescribed by Gross (104); *Asterosteus* and *Jagorina* were very poorly known until reinvestigated by Stensiö (97); *Radotina* and *Kosoraspis* from the very base of the Devonian in Bohemia (not Silurian, as first thought) are described by Gross (105); *Kolymaspis* was described by Bystrow in 1956. In *Gemuendina,* where alone the complete body is known, we are dealing with a form with a depressed body and pectoral fins greatly expanded to give a skatelike appearance. The armor is much reduced, but as Watson first clearly pointed out, despite reduction, we are dealing with armor of the basic placoderm type, with the characteristic division into "cranial" and body segments and a pectoral spine which is persistent despite fin expansion. The "skull" pattern differs in various ways from that of typical arthrodires, notably in such features as the medially placed eyes and the presence of a pair of dorsally (and medially) placed nostrils. *Radotina, Kosoraspis, Kolymaspis,* and *Palaeacanthaspis* are less completely known, but appear to be other early members of the same group; *Jagorina* and *Asterosteus* are later Devonian survivors. That these rhenanids are related to the general arthrodire group is clear. But as is equally clear from their structure, they must have separated from the more "normal" arthrodires well back in the Silurian, at least, and it is of interest that they (and the petalichthyids) are marine in habitat, whereas the Lower Devonian normal arthrodires were mainly fresh water.

The ancestral placoderm?—When we first see the placoderms at the base of the Devonian, they are already a highly diversified group, including rhenanids, petalichthyids, and arctolepid arthrodires, and perhaps even antiarchs. There obviously lay behind them a long evolutionary history of which we know absolutely nothing, and we can do no more than speculate about what the Silurian "ur-placoderm" would have been like. Surely there was an extensive head and body covering of the typical placoderm type, and, as suggested by the presence of scales in forms as far apart as *Lunaspis* and *Astrolepis*, the non-armored regions of the body were completely scale covered. Again, well-developed pectoral spines would have been present, with the paired fins small, at the most. Still further, the depressed shape of many of the "funny-fishes," as well as early arthrodires, suggest that the ancestors were flattened bottom dwellers. We can say nothing as to jaw structure, but presumably jaws were weak. And, were it not absolute heresy, I might fancy that the remote ancestor was, rather like the cephalaspids, essentially a mud-grubber, in which the development of biting jaws was barely begun—if begun.

Shark ancestry.—With the acceptance of the thesis of skeletal degeneration in many groups of fishes has come the possibility that the elasmobranchs are secondarily boneless and descended from bony ancestors. They surely are not descended from any member of the Osteichthyes proper, and descent from the Acanthodii is improbable. This leaves us with the Placodermi as the only remaining gnathostome group from which the sharks might have been derived. There is no inherent impossibility in this, and Stensiö, who has long advocated this pedigree for the elasmobranchs, is so confident that he would unite them with the placoderms in a superclass Elasmobranchiomorphi.

But from which group of placoderms can the sharks have been derived? There is no particular indication of close resemblance to the shark pattern in any of the placoderm groups reviewed. Further, nearly all the known groups, while showing frequent reduction in the extent of body armor, show no strong signs of major degeneration of the armor retained. There is, however, some suggestion of armor reduction in the *Stensioella* group of petalichthyids, and very definite armor reduction in *Gemuendina* and its rhenanid relatives. Are we finding clues here to shark origins? In the rhenanids the peculiar position of the eyes and nostrils and (in *Gemuendina*, at least) the exuberant development of the pectoral fins are among structural features suggesting that we are at the most dealing with an analogue to the line of descent taken by the elasmobranchs and certainly not with direct ancestors. The petalichthyids are less unorthodox in structure, and although *Macropetalichthys* and its close relatives can be ruled out, it is possible that relationships may exist between the *Stensioella* group and potential shark ancestors.

Smith Woodward (1924) long ago described as *Cratoselache* a form from the marine Lower Carboniferous of Belgium which suggests a sharklike form with degenerate remnants of placoderm plates. The period is, of course, too late for *Cratoselache* to be in itself an ancestor of the sharks—then already flourishing—but it is not impossible that it is a late survivor of a transitional form.

Palaeospondylus.—In earlier decades much was written about the probable nature of this little organism, found abundantly in a single Scottish quarry. Nothing new has been added, since Moy-Thomas's paper of 1940 (112), to either knowledge of it or debate concerning it, and it still remains a "problem child."

Acanthodians.—The acanthodians, the so-called "spiny sharks," are by far the oldest known of gnathostome vertebrates (no placoderm, even, is known below the Silurian-Devonian boundary, as noted above, and other gnathostome fishes are still later in appearance). Articulated acanthodian skeletons are present only in the early Devonian and later times, but there are fairly frequent reports of scales and spines of this group well back into the Silurian (cf., for example, Squirrel 1958; Gross 1947). Some early workers thought them "ganoids"—i.e., Osteichthyes—but since a time well back in the last century a commonly held view has been that, despite the presence of a bony skeleton and good bony scales, they are in some way related to the elasmobranchs. Important in scientific work on acanthodians was Watson's monograph of 1937 (123), in which he was able to give considerable morphological detail. Watson believed that the acanthodians were very primitive in not having modified the hyoid arch as a jaw prop, and hence retaining (as theoretically one would expect of ancestral gnathostomes) a full hyoidean gill slit. He believed that the placoderms also retained this primitive "aphetohyoidean" condition, and considered placoderms and acanthodians to constitute a common group of archaic gnathostomes. It was apparent, however, that apart from this supposed common primitive feature, there was little to unite acanthodians and placoderms, and they have certainly had a long separate history (cf., for example, Westoll 88, Fig. 9; and Denison 92). Watson's contention that the acanthodians had an unspecialized hyoid and spiracular region was vigorously contested by Stensiö (1947) and others, and as is further shown by the work of Miles mentioned below, the hyomandibular apparently functions as an aid in jaw support, as in many characteristic jawed fishes; if the diagrammatic primitive "aphetohyoidean" gnathostome ever existed, it must have been at a lower level (and an earlier one) in the history of jaw evolution.

In opposition to earlier beliefs as to the elasmobranch affinities of the acanthodians, pieces of evidence have gradually accumulated suggesting some type of relationship to the Osteichthyes and, curiously, several features specifically suggesting the actinopterygians. It has long been known, for example, that acanthodian scales resemble, structurally, those of palaeoniscoids; Heyler (1958, 1962) has commented on the similarity of the caudal fin of acanthodians to that of palaeoniscoids; the acanthodian head, with large orbits and a short snout, is comparable to that of early actinopterygians, and suggests that the acanthodians were similarly "eye-fishes," depending more on vision than smell for sensory apperception. Miles (124) has recently begun a restudy of acanthodian anatomy with interesting results. As he notes, it is unfortunate that internal structure is little known, except in the terminal Permian form *Acanthodes*, and regrettable that we know little of the *Climatius* group, which may have differed considerably from other acanthodians. Although discounting specific resemblances to actinopterygians as probably due to parallelism, Miles notes many features in which the acanthodians are basically similar to the more generalized Osteichthyes, features such as the structure and position of the hyomandibular of the amphistylic (not autostylic) jaw, the build of the gill bars, the development of the spiracular tube, the presence of a large hyoidean gill cover in most, at least, of the acanthodians, the presence of the typical osteichthyan basal articulation between braincase and palatoquadrate, and the development of large, compact otoliths. Definitely, the acanthodians are related to the bony fishes. But, on the other hand, they are considerably farther removed in structure from the "normal" osteichthyan groups—actinopterygians, crossopterygians, lungfishes—than members of these three groups are from one another. In contrast to them, the acanthodians lack, for example, dermal palatal and marginal jaw bones, further lack the pattern of dermal shoulder girdle structure common to the three, and there is no reason to believe that the spiny structure of the paired fins is other than a very divergent specialization of the acanthodians. Although it seems reasonable to believe that the acanthodians and typical osteichthyans are descended from a common remote ancestor, the divergence of the acanthodians from the common stock presumably took place at an earlier time than that at which the other groups separated from one another.

How to array the acanthodians and the more typical bony fishes in a formal classification is troublesome. One may consider that acanthodian divergence is so great that they should be placed in a discrete class or other major category. But the resemblances to the bony fishes seem undeniable, and some sort of bracketing with them seems reasonable. Miles suggests that instead of the three classes into which gnathostome fishes are frequently assembled, the three customary osteichthyan subclasses be elevated to class level, and the acanthodians, as a fourth class, be united with them in a superclass. I strongly disagree with this suggestion. There is no need for elevation of the bony fish subgroups to class level, and the consequent necessity of calling the broader group a superclass. It is simpler and more sensible, I think, to retain the Osteichthyes as a class, including (perhaps with a question mark) the acanthodians as well as the usual components of the group at a subclass level.

4

Sharklike Fishes

In line with earlier beliefs about skeletal history, it was but natural in earlier days that the jawed fishes, above the cyclostome level, should be divided into two groups, a supposedly inferior one in which the skeleton was unossified, and a supposedly advanced group with an ossified skeleton. The term Chondrichthyes has been frequently and appropriately applied to the former group, in which are classed the sharks, skates and rays, and chimaeras. The group was presumed to be a natural one; in addition to the absence of bone, sharks, on the one hand, and chimaeras, on the other, possessed such common distinctive features, otherwise unknown in fishes, as pelvic claspers for mating purposes and large-yolked eggs.

Currently, however, this seemingly simple and obvious method of classification is in serious difficulties. As we have discussed earlier, the work of Stensiö, in particular, strongly suggests (although it does not clearly prove) that the sharks are descended, with skeletal degeneration, from placoderms of some sort; further, that the chimaeras are also descended from placoderms, and probably from placoderms of a totally different subgroup. If this diphyletic origin of the two major components of the Chondrichthyes from the Placodermi is confirmed, the class Chondrichthyes will prove to be an unnatural assemblage; two derived types should not be classed together without including their common ancestors. Hence it will be necessary to abandon the Chondrichthyes as a unit in classification and combine sharks and chimaeras with the placoderms in one vast heterogeneous assemblage almost impossible to define or characterize. Stensiö in all his more recent works has taken this step, terming this supergroup the Elasmobranchiomorphi. I have not done so in the present edition of my work, pleading the excuse that the case for shark and holocephalian origins is not proved. But I fear that my conservatism here is unjustified, and that future discoveries may give evidence making such a fusion necessary.

The Chondrichthyes as currently constituted are clearly divisible into two units—on the one hand, the Chimaerae, or Holocephali (discussed later), on the other, the sharks and forms such as skates and rays derived from them. There is some question as to terminology, but it seems best to use Elasmobranchii (sometimes utilized in a narrower sense) to include the whole shark-ray assemblage, which seems to be a natural group.

Early sharks.—Elasmobranch history can, in general, be divided into three stages: (1) the primitive cladoselachian stage characteristic of the Devonian and presumably lingering on into the later Paleozoic; (2) a hybodont stage, which appears to have begun in the Mississippian, and dominated until the Triassic; (3) a modern stage, starting in the Jurassic with the appearance and radiation of modern shark types and, presently, of their skate and ray relatives.

In line with earlier beliefs about the relatively primitive nature of cartilaginous fishes, one would have expected the sharks to have appeared early in the fossil record. But, as is well known, this is the reverse of the case. Except for isolated teeth in the Middle Devonian limestones of Ohio, there is no trace of a shark until the late Devonian, whereas earlier in that period we have, of gnathostome fishes, not only a wealth of placoderms, but abundant representatives of the major groups of "higher" bony fishes. Our inability to trace them farther back is puzzling. In contrast to some other groups, the elasmobranchs are dominantly marine in habitus (the pleuracanths are the only exception of importance), but there is no trace of a shark in the widespread marine rocks of the earlier Devonian. An appeal to the old belief in the primitively cartilaginous condition of the skeleton does not solve the problem, for even if one were to assume the ancestral sharks to have had skeletons lacking calcification, and hence unlikely to be preserved, we should at least find teeth (as we do in abundance later). Under current prevalent belief that the sharks are skeletally degenerate placoderms, we would reasonably expect to find earlier forms transitional from placoderms. But we do not.

Although fragmentary remains (teeth and a few braincases) have been found elsewhere, our best picture of early elasmobranchs is obtained from the black Cleveland Shale of late Devonian age, in which a peculiar type of environment has preserved not merely calcified materials but sometimes soft tissues, and in which is present *Cladoselache*, the presumed archetypal shark.

In earlier days, many specimens were recovered from the Cleveland Shale, but little new data have been published recently except for a new restoration by Harris (1951), in which he points out that some specimens, at least, carried a dorsal fin spine. New interstate road construction in the Cleveland region is currently yielding large quantities of Cleveland Shale fish material, which one may hope will give us much new information concerning late Devonian sharks (as well as contemporary arthrodires). In *Cladoselache*, claspers appear not to have been developed, and the paired fins are broad based and of a supposedly very primitive type. Succeeding the cladoselachians in the later Paleozoic were the hybodonts, in which the primitive amphistylic jaw suspension was retained; the fins, however, were modified toward and to the narrow-based typical shark type and claspers were present. Presumably the old-fashioned cladoselachians overlapped the hybodonts into the Carboniferous (or even Permian), but one would think that it would be relatively simple to sort the two types out. Not so; there are numerous difficulties caused, among other things, by the fact that late Paleozoic shark remains are generally fragmentary in nature, often consisting only of teeth and spines. The teeth of *Cladoselache* are of a type to which the name *Cladodus* was early applied; such teeth are common in the Carboniferous, and while some forms bearing them may well be cladoselachian, some are definitely associated with hybodont forms. Complicating the picture is the *Ctenacanthus* problem. This name was given originally to a characteristic Carboniferous spine type. Associated materials show that sharks with *Ctenacanthus* spines had teeth identical with the *Cladodus* type; and, further, while Carboniferous evidence shows *Ctenacanthus* spine-bearing sharks to have paired fins of hybodont pattern, specimens from the Cleveland Shale with *Ctenacanthus* spines show fins (as well as teeth) identical with those of *Cladoselache*. As if this did not make enough confusion, it may be noted that although *Cladoselache* and the Devonian "*Ctenacanthus*" lack claspers, such structures are already present in a rare shark, *Diademodus* (Harris 1951), from the Cleveland Shale, as well as in at least some specimens of *Ctenacanthus* (*sensu stricto*) in the Carboniferous. Altogether, much needs to be learned before early shark history can be satisfactorily unraveled.

Until recently, little has been known of the braincase structure of early sharks. In modern elasmobranchs the postorbital position of the braincase is extremely short, and it was tacitly assumed that the same was true in older forms. Hussakof (1911) had described braincase specimens of the Paleozoic pleuracanth sharks on the same assumption; unfortunately, however, his specimens were incomplete, so that his interpretation of regions along the longitudinal axis is inaccurate, and, in addition, he believed the true dorsal surface to be ventral, and vice versa. Many years ago I found excellent pleuracanth cranial materials in the Texas Permian; sketches have been given of these in successive editions of my *Vertebrate Paleontology* but, regrettably, I have never published a full description of this material. Notable is the great length of the otic region, in strong contrast with modern sharks.

Both Stensiö (1937) and Gross (1937) have described crania of Lower Carboniferous cladodont sharks. As far as their specimens were preserved, they agreed well with the pleuracanth material in structural features. In neither case, however, was the posterior end of the braincase complete, and they hence reasonably interpreted their specimens as having short otico-occipital regions. I have recently been able to supplement their findings by the description (118) of a further early elasmobranch, *Tamiobatis*. This was thought by its original describer, Eastman, to be a Paleozoic skate, with one surface—actually ventral, but thought by him to be dorsal—exhibited in a limestone nodule. This specimen, I found, could readily be prepared by acid treatment. There is no dentition associated, but since the environment, as shown by the matrix, was obviously marine rather than fresh water, it can reasonably be interpreted as that of a true elasmobranch, rather than a pleuracanth. It agrees well both with the cladodont specimens of Stensiö and Gross and with the pleuracanth braincase. A further Mississippian marine braincase will be described by Dunkle. In both of our specimens there is, as in pleuracanths, a very long otico-occipital region, suggesting—although not proving—that the proportions seen here were those common in early sharklike fishes. The proportions, compared with those of later types, suggest an analogy with the shift in skull proportions between the rhipidistian crossopterygians and their tetrapod descendants, in which series the posterior part of the cranium becomes markedly decreased in length in relation to the anterior region. In view of the theory that sharks are descended from placoderms, comparison of these early shark braincases with those of placoderms, studied especially by Stensiö (98, 1963), is indicated. In none of the placoderms in which the braincase is well known do we find any close comparison with the proportions found in the known older sharks; all are much broader, for example, and while in some, at least, the proportion of preorbital to postorbital length is rather lower than in modern sharks, in none do we see an otic region as greatly elongated, relatively, as in the known Paleozoic specimens.

I have available a considerable series of pleuracanth braincase sections made by the peel method; Dunkle's specimen was cut, in the quarrying operations which discovered it, into a number of cross-sections. It is to be hoped that we may presently be able to give an account of the internal anatomy of the Paleozoic shark braincase.

Hybodonts.—The Mesozoic genus *Hybodus* is typical of a group of sharks which appear to have evolved from and to have succeeded the cladoselachians, retaining the seemingly primitive amphistylic jaw suspension but gaining a narrow-based fin and developing claspers (and, further, having a short braincase). Surely, as shown by a few well-preserved specimens, the hybodont type was already developed in the Carboniferous, but, as noted above, the situation is confused by an overlap in spine and tooth types with the cladoselachians, and by the fact that most Carboniferous finds of sharks consist merely of teeth or spines. Much more data are needed regarding the nature of Palaeozoic hybodonts. Current studies of the fauna of Pennsylvanian black shales by Zangerl (cf. Zangerl and Richardson 1963) will, it is hoped, shed some light on this confused situation. Typical hybodonts have a heterodont dentition, with sharp teeth, tending to be retained in spiral series, in the front of the mouth and, farther back, broadened multicusped teeth suitable for dealing with "shellfish." This "all purpose" dentition may have been responsible for enabling the hybodonts to survive the "hard times" at the end of the Paleozoic. In any discussion of elasmobranch phylogeny, it must be remembered that no type of shark other than hybodont is known from the Triassic, and that, hence, on the basis of present knowledge, all later types of sharks (and rays) must be assumed to have been of hybodont origin. Suggestions, such as a recent one by Jarvik (1965*c*), that skates were derived from rhenanids is grotesquely improbable when we consider, among other factors, a time gap of nearly five full periods intervening without any trace of skatelike types!

Characteristic of the Carboniferous and persisting into the Permian were members of the Edestidae, in which the cheek teeth were highly developed and in which a curious feature was the retention of large teeth in the symphysial series to form a complex spiral, seen at its height in *Helicoprion* of the Permian; among the more recent workers publishing on edestids have been Nielsen (1952), Obruchev (1953), and Bendix-Almgreen (1966); Liu and Chang (1963) have recently described a new edestid, *Sinohelicoprion*, from China. Zangerl (1966) has described an edestid, *Ornithoprion*, of grotesque structure, and reports the presence of material of several others in Pennsylvanian black shales.

In edestids the biting surface of the cheek teeth has a columnar structure of the dentine. This is similar to the histological structure seen in the teeth of bradyodonts, and Nielsen has claimed that in consequence, the edestids should be placed in the bradyodont assemblage. But in all other regards, the edestids seem to be merely somewhat specialized hybodonts, and the tooth histological pattern (obviously associated with the development of a good crushing surface) may have developed in parallel fashion (Radinsky 1961; Ørvig 1965*b*). Bendix-Almgreen concedes that edestids are shark (rather than chimaera) relatives, but would separate them ordinally—on, I think, insufficient grounds.

Not uncommon in the Cretaceous are flattened button-like teeth described principally as *Ptychodus*. These teeth were obviously arrayed in a pavement indicating molluscivorous habits. The ptychodonts show no indication of relationship to either the ancient bradyodonts or to typical skates and rays, and their position was long a puzzle. A recent excellent monograph by Casier (119) has shown that they are a hybodontoid side branch.

"Modern" elasmobranchs.—In the Jurassic we enter the realm of sharks, skates, and rays of modern types, and from the Cretaceous onward many of the living genera are present and are common (mainly represented by teeth) in numerous deposits. As in earlier decades, faunal studies of various marine deposits of the late Mesozoic and Tertiary have yielded considerable shark material—Arambourg and Signeux (53), Casier (55), Glikman (1958, 1962), Gurr (1963), Schaeffer (1963), and Tabaste (1964) may be cited among those describing specimens of this sort; Casier (114) has broadly reviewed much of the data on fossil shark distribution. Identification is hampered by nomenclatorial problems in many cases; the works of Bigelow and Schroeder (115), although dealing only with existing forms, are invaluable in straightening out nomenclature and in giving excellent illustrations of dentitions. It is notable that considerable variation may occur in the teeth present in a single jaw, and it is regrettable that certainly in some cases and probably in many others, several specific names have been given to tooth types that could have been present in a single individual.

Pleuracanths.—The one notable exception to the general statement that sharks are marine is found in the Pleuracanthodii, common in freshwater deposits in the Carboniferous and early Permian. Only in the Bohemian Gaskohle faunas described by Fritsch is the entire body preserved. Although surely elasmobranchs in origin, they are highly divergent in fin types, teeth, and the presence of a large occipital spine. In their cranial anatomy, however, they appear to agree well with primitive cladoselachians; I have in possession excellent material of braincase and jaws (apparently similar to those of other early elasmobranchs) which I hope to describe in the near future. Their early and late history is obscure. Presumably they represent invaders into fresh water from the sea, but there is little known of them before the Pennsylvanian. Abundant at that period and in the early Permian, they are unknown in the later Permian and most of the Triassic; but they are definitely present in beds in Australia said to be late Triassic, and there is some slight evidence of their presence in Europe at this time. Their absence in the intervening stages is puzzling, but may be due to the absence of beds deposited under proper environmental conditions.

Bradyodonts.—Highly specialized for the treatment of molluscs and other shelled invertebrates were the various families of "pavement-toothed sharks," which made their appearance at the end of the Devonian, flourished in the Carboniferous, and survived into the Permian—Petalodontidae, Copodontidae, Helodontidae, Psammodontidae, Chondrenchelyidae, and Cochliodontidae. How to classify them was—and is—a major problem. Smith Woodward in 1921 suggested that they be placed in a common group as the Bradyodonti; I seized upon this solution with pleasure, and used this terminology in earlier editions of my work. It is, however, obvious that this assemblage is an artificial one; there is no evidence that any two of the families concerned are related to one another, and surely all have arisen independently from primitive elasmobranch——or pre-elasmobranch—ancestors. As noted later, the cochliodonts may be related to the chimaeras, and will be mentioned in that connection. As regards the other families, except in a very few cases we know little of them except for their varied tooth plates, a situation which adds to the difficulty of attempting to determine their relationships. All five of the families here considered were flourishing in the Mississippian, and hence their origins must be sought in the Devonian (and indeed two, possibly three, of these families are reported from the late Devonian); but we have no Devonian shark ancestors from which to derive them except the seemingly primitive (and rather late) cladoselachians, and there are no known placoderms suggestive of being antecedent to them. This is a very puzzling chapter in fish history.

Chimaeras.—Apart from egg type and claspers (and the presumably secondary absence of bone), little ties the chimaeras to the shark group. Not much has been done regarding fossil chimaerids lately until Patterson's (122*a*) recent excellent review. Their presence in a variety of forms in the Jurassic contrasts strongly with the apparent complete absence of chimaeras or possible antecedent forms in the Triassic.[1]

We have already noted the theory of the descent of chimaeras from ptyctodonts. Patterson brings evidence of an alternative theory of their origin from arthrodires, but from a non-ptyctodont source; I have, after some hesitation, adopted Patterson's point of view rather than the theory of ptyctodont relationships. It has often been noted that members of the cochliodont family of late Paleozoic "bradyodonts" have tooth plates suggestive of those of chimaeras (and of ptyctodonts). Patterson points out that the last surviving cochliodont, *Menaspis* of the late Permian, shows remains of a degenerate armor and has "head processes" somewhat comparable to those of later chimaeras, and that while most other cochliodonts are known from teeth alone, *Deltoptychius* has dorsal bony plates which presumably are remnants of a once extensive dermal armor. These forms suggest, as does the ptyctodont theory, a descent of chimaeras from arthrodires. But the body plates found in the cochliodonts mentioned are not at all of ptyctodont pattern and imply chimaeroid origins from some unfamiliar arthrodire type.

Some points in Patterson's broader conclusions, however, I find puzzling. He dissociates the forms mentioned from the other cochliodonts, apparently with no reason other than the negative one that no armor has been found associated with other cochliodont genera; and, on the other hand, purely on the basis of tooth histology (*contra* Radinsky and Ørvig), he associates broadly with the Holocephali not only cochliodonts, but the whole sweep of Paleozoic "sharks" with pavement teeth—even the edestids, which seem rather surely of elasmobranch descent. I have, in the classification in my text, included the cochliodonts *in toto* among the Holocephali, but refused to include other "bradyodonts" or edestids.

"Ichthyodorulites."—In the Carboniferous (and to a lesser

[1] Curiously, in 1933 both Norman, in his *History of Fishes*, and I, in the first edition of *Vertebrate Paleontology*, independently of one another implied the presence of at least fragmentary Triassic remains of chimaeras; but when later questioned, neither of us was able to locate data supporting this.

extent in the Devonian), there have been described numerous spines, isolated and not associated even with dentition, which presumably pertain to lower fish types. Over the years a number of these have been linked with fish of one sort or another, but the significance of some two dozen or more remaining orphan "ichthyodorulites" still perplexes us.

5

Bony Fishes

Earlier beliefs as to skeletal history naturally assumed that the existing types of bony fishes and their fossil relatives formed a natural group of "higher" fish types. Despite changes in modern points of view, there is still no reason to doubt that these forms—actinopterygians, crossopterygians, and lungfishes—are a phyletic unit, derived ultimately from a common ancestor. To them the term Osteichthyes is commonly applied and may well be retained, despite the fact that bone was early present in other fish groups. As discussed earlier, it is probable that the acanthodians are related to them, although this relationship is obviously a remote one.

In some recent publications (Jarvik 8; Stensiö 98; Miles 124) the word "Teleostomi" has been used as a substitute for Osteichthyes. This usage is contradictory to the meaning and proper use of "Teleostomi" and should be avoided. The name ("terminal mouth") was coined by Owen and long widely used (cf., for example, Goodrich 31) to include actinopterygians and crossopterygians and definitely to *exclude* the lungfishes, with an underslung, non-terminal mouth. Obviously it should not be used as a synonym for Osteichthyes.

Within the bony fishes it has long seemed clear to most workers that three major groups are present (the case of *Polypterus* is noted later). These are the Actinopterygii, or ray-finned fishes, the Crossopterygii, and the Dipnoi, or lungfishes. What are the interrelationships of the three? Some workers in the past and some today, such as Stensiö and his school and White (1966), believe that all three are equally discrete from one another. On the other hand, a long series of students of fishes, such as Goodrich, Gregory, and Watson, particularly, have pointed out that despite obvious differences in the skeleton, the crossopterygians and dipnoans exhibit characters strongly suggesting that they are more closely related to one another than either is to the ray-finned forms. This is not so apparent if we consider, as paleontologists, only skeletal structures, for the braincase and dentition differ strongly in even the earliest members of both groups. But even as regards data available from the fossils, there are notable points of similarity and of differences from the actinopterygians. The scales of early crossopterygians and dipnoans are of the cosmoid type, and differ sharply from the ganoid scales of ray-finned forms; and the paired fins are diagnostic. Only exceptionally is there any extension of skeleton and muscle into the fins in ray-finned forms, where the skeletal fin supports are typically short parallel rods, and the main bulk of the fin is formed (as the name implies) by a web of skin supported by fin-rays. In strong contrast, flesh and skeleton extend far out into the fin in both crossopterygians and dipnoans, and the fin skeleton, in contrast with actinopterygians, is of an archipterygial type—sometimes typical, sometimes abbreviate—with a central axis and side branches.

There is thus considerable skeletal evidence to support crossopterygian–dipnoan relationship, despite the marked differences in cranial structures. But even the paleontologist should take a broader view of the question than that which can be derived solely from the skeleton. The skeleton is, of course, but one of a long series of organ systems which should be brought into consideration. What of these other systems? What of the mode of development?

Here the evidence is clear and positive. In the lungfishes, many anatomical features of the "soft anatomy" are highly comparable to those of modern amphibians—so much so that the Dipnoi were long thought to be tetrapod ancestors—and contrast strongly with those of actinopterygians. Further, the mode of development of the lungfish egg and larva is very similar to that of amphibians, whereas the actinopterygians have a highly specialized developmental pattern The evidence about the ancestral crossopterygians is indirect. We cannot, of course, ever expect to gain much knowledge of the soft anatomy of the rhipidistians of the Paleozoic, and the only surviving member of the Crossopterygii, *Latimeria*, is obviously so far removed from the ancient central stock that we would not expect it to be too representative of the original pattern. But since it is, I think, universally agreed that the amphibians are descended from rhipidistians, or at least forms closely related to them, one can say with confidence that the soft anatomy and reproductive patterns of these ancient crossopterygians were similar to those of their amphibian descendants and equally close to those of the Dipnoi.

I think, thus, that any broad consideration of the problem leads to the conclusion that there are basically two major groups within the Osteichthyes—one, the Actinopterygii; the second, the Crossopterygii and Dipnoi combined. The two may reasonably be considered as subclasses, and a name is needed for the second. Hubbs (1919) long ago suggested Amphibioidei; this, however, is not only a rather awkward term to use, but is also not too descriptive, since it merely suggests one of many evolutionary lines to which the ancestral members of the group gave rise. In 1937 I suggested Choanichthyes as more appropriate (Romer 148), on the then prevalent assumption that internal nares, or choanae, were present in lungfish and all crossopterygians. More recent work seems to show that the internal nostrils of lungfishes are not true choanae, however, and that choanae are not present in all crossopterygians. I therefore suggested (Romer, 1955*a*) that Sarcopterygii, "fleshy-finned fishes," would be an appropriate and didactically useful term, in contrast to Actinopterygii. This terminology has been fairly generally accepted. However, Dr. Trewavas and other British Museum workers (1955) countered with the proposal that, instead, the term Crossopterygii be broadened to include the Dipnoi as well as the Crossopterygii in the usual sense of that term. Apparently such usage had been common within the British Museum, but at that time unknown beyond its portals. I trust that there will be no general acceptance of such an aberrant usage of "Crossoptergyii"; if this were done, it would throw the entire literature of the subject into hopeless confusion.

All three major types of bony fishes appear, fully characterized and distinct from one another, early in the Devonian. Whence did they come? For one who believes in the late development of bone and, as well, in marine vertebrate origins, it may be fancied that all three types first appear in the fossil record at this time because they had just acquired, independently, calcified bony structures. This I find difficult to accept. The alternative—to me more probable—is that there was a long antecedent history of forms with a bony skeleton, which lived in some other environment. This environment, I personally think, was in fresh waters; and almost no truly continental sediments are present in the geological record antecedent to the Devonian and very late Silurian redbed deposits. There are—and presumably already were in the Devonian—notable "soft anatomy" differences between the ray-finned and fleshy-finned groups, and hence the point of departure must have been well antecedent to the time of their appearance. In consequence, it may be suggested that we are not dealing, in the Osteichthyes, with a coherent group, but with two or three discrete branches from a generalized primitive gnathostome stock. But the three typical osteichthyan groups, plus the acanthodians, seem to have a similar basic structural pattern, and one which differs notably from that of the only other truly ancient gnathostome pattern, present in placoderms. Presumably, still farther back, placoderms and osteichthyans converged; but this point of divergence must have been very remote indeed.

Crossopterygians.—The rhipidistians are the prominent Paleozoic members of the Crossopterygii, appearing in flourishing condition in mid-Devonian fresh waters and surviving to the early Permian. Rare, however, at first, and apparently tending strongly toward a marine habitus, were the Coelacanthini (or Actinistia), which in contrast to the rhipidistians have survived in the form of *Latimeria.* Some authors (Stensiö 1963, for example) would consider the differences between rhipidistians and coelacanths marked enough in such features as the absence of choanae in coelacanths, so that the two should be separated as widely from one another as either is from the Dipnoi and Actinopterygii (even to erecting separate classes for them). This does not seem merited; the fact that the two types possess in common the highly specialized dual structure of the braincase, markedly in contrast to that of any other fishes, is in itself sufficient for associating them.

The rhipidistians have attracted the attention of morphologists for a century; Goodrich, Watson, and Gregory are among the workers who in the earlier part of the present century made careful studies of the more superficial features of their anatomy. The past two decades have seen the publication of a long series of papers describing fossil rhipidistians, notably by Jarvik (147, 1950*a*), but also by Gross (149) and Vorobjeva (1959, 1962). In recent decades we have come to know much of the internal structures as well. I contributed to a minor extent with a description of the braincase of "*Megalichthys*" (*Ectosteorhachis*) (148), but we owe especially to Jarvik (152, 154) a series of splendid reconstructions of the anatomy of rhipidistians. Especially interesting is the fact that the braincase is in two discrete and articulated pieces, through the posterior of which a large round canal extends ventrally. So remarkable and so seemingly specialized is this structure that it long appeared to be a stumbling block in consideration of the rhipidistians as tetrapod ancestors; this difficulty, however, has been happily removed by the discovery that the canal structure was still present in the oldest known amphibians.[1] The ventral canal was originally suggested (Stensiö 1932) to have carried musculature. My studies of "*Megalichthys*" led me to believe that it was, on the contrary, occupied by an unrestricted continuation of the notochord. At the time it seemed highly improbable that the question could ever be resolved. What was my delight, after the discovery of *Latimeria*, to see dissected specimens with this canal snugly filled by the notochord!

Because of presumed ancestry to that of amphibians, the dermal skull pattern of crossopterygians has long attracted much attention. As Westoll and I, among others, have suggested, it appears probable that the dermal skull roof of ancestral bony fishes was probably a variable mosaic of small bones. Among the lungfish the mosaic persisted, although with an eventual trend toward reduction in numbers of elements and toward stabilization. In crossopterygians the trend toward reduction and stabilization was stronger and faster, so that in Devonian forms the posterior part of the skull had achieved an essentially fixed pattern of large elements, although the anterior

[1] The history of this braincase structure is an excellent example of the fact that we should not assume, as is often done, that "specializations" cannot be eradicated in phylogeny. The braincase in all typical tetrapods is a unit structure, and it presumably was also a unit in an ancestral gnathostome; but contrary to what one would tend to believe, part way along the story there evolved, only to be again eliminated, this highly specialized crossopterygian structure.

area still consisted of numerous small bones which exhibited considerable variation. Certain elements in the rhipidistian pattern were obviously directly comparable to those of tetrapods, but in other cases there were problems. Notable for years was the puzzle concerning the position of the parietal foramen for the "third eye." As discussed in the next chapter, this problem and the identification of the crossopterygian parietals and frontals has been resolved.

Much has been written concerning the paired-fin structure of rhipidistians—particularly the pectoral fin. As in dipnoans, some rhipidistians have a full-fledged archipterygium, but most have an abbreviated form of the same pattern. Certainly this fin type is of the sort from which the tetrapod limb can be derived. But rather surely no known genus lies directly on the ancestral line of the Amphibia, and I feel none too sure of the success of attempts to make detailed comparisons between the distal elements (surely variable) of the rhipidistian fin and the tetrapod manus.

In connection with his theory of the dual origin of tetrapods (discussed in the next chapter), Jarvik has in various works pointed out anatomical contrasts between two groups of rhipidistians, represented by *Porolepis* and *Holoptychus*, on the one hand, and *Osteolepis*, *Eusthenopteron*, etc., on the other, and has considered the differences sufficient to erect here two major groups, Porolepiformes and Osteolepiformes, which in such a classification as Stensiö's (1963), are raised to class level. This does not seem warranted to me. Differences do exist; but as far as described do not seem to be of a magnitude greater than that which we normally find within the limits of superfamily or suborder. Of the two groups, that including *Holoptychus* and *Porolepis* had been in great measure neglected in earlier times, and we are greatly indebted to Jarvik for extending our acquaintance of it. Rather than use the name *Porolepis*, an inadequately known form, I prefer to use the familiar name *Holoptychus* for this group, as the Holoptychoidea. Of the other subgroup of rhipidistians, the structure of *Eusthenopteron* is now very adequately known, and we have considerable knowledge of such other forms as *Ectosteorhachis*, *Megalichthys*, and *Osteolepis*. They may reasonably be known as the Osteolepidoidea. Clearly distinguishable here are Osteolepidae and the more predaceous Rhizodontidae. There were large Carboniferous members of the latter family, such as *Rhizodus*, which are still inadequately known and are being studied by Thomson.

Onychodus, known mainly from tooth whorls from marine Devonian (mainly Middle Devonian) deposits in various regions of the world, long was puzzling as to its relationships. Better knowledge of the Holoptychidae through Jarvik's work indicated that *Onychodus* was a related type. Ørvig (41) stated that it could be included in that family; I felt, however, that it might well represent a separate family.

After my volume went to press, there appeared the portion of the Piveteau Traité (1966) dealing with the bony fishes. In this was inserted a section by H. L. Jessen (pp. 387–98) describing a new, tiny Devonian crossopterygian, *Strunius*, which, its describer points out, belongs in the same family as *Onychodus*. Jessen has since published more fully on this material (1966). For *Strunius* and *Onychodus* Jessen proposed the erection of a new major group of crossopterygians, the order Struniiformes, ranking with the Rhipidistia and Coelacanthini. As Lehman, the translator of his article, suggests, this proposal is, to say the least, premature, and I was unable to consider it in the text of my work, although inserting the generic name *Strunius* in the classification in the family Onychodontidae.

How great the differences between *Strunius-Onychodus* and the *Holoptychus-Porolepis* group of rhipidistians are is a matter worthy of consideration. But it does not seem probable that they are of a sort meriting ordinal separation; further, it would seem better, if a new group were to be named, to create one formed on the well-known generic term *Onychodus* rather than on this newly described genus.

Coelacanths.—Never abundant and certainly, in their later stages at least, purely marine in nature, the coelacanths, as definitely a minor side branch of the crossopterygian stock, have, until recently, attracted relatively little attention. Stensiö, however, early interested himself in the group, particularly as regards the braincase in early forms, and Schaeffer has recently contributed as well (Schaeffer 1952*a*, 1952*b*; cf. also Lehman 46; Gardiner 138). Highly important, of course, has been the discovery of *Latimeria* and the careful descriptions by Millot and Anthony of this fish's anatomy, which give us our only available clue (if a rather faint one) about the probable "soft anatomy" of the ancestral crossopterygians. A volume on skeleton and muscles appeared in 1958 (158) and a second volume on the nerves and sense organs, in 1965 (159).

Dipnoans.—Dollo (164) first established the general picture of the evolution of body and fin form along the main line of lungfish evolution; in the last few decades a number of new types, including aberrant long-snouted forms (Graham-Smith and Westoll 169; Säve-Söderbergh 1937), have broadened the picture. For many years the pattern of skull bones was a center of interest, and many attempts at interpretation of this mosaic were made by a variety of authors, such as Forster-Cooper, myself (1936, with poor results), Parrington (1950), and White (168). Westoll (165) in 1949 gave an excellent resumé of dipnoan evolution as known up to that date; there have been in recent years a number of descriptions of primitive Devonian forms by Gross (149), Säve-Söderbergh (167), Ørvig (1960), and White (1962). Lehman (170) has described a new series of Devonian forms from Greenland. In lungfishes, it appears, internal skeletal degeneration proceeded rapidly, so that in most forms we know nothing of internal cranial anatomy; recently, however, endocranial remains have on occasion been found in Devonian forms and have been described in such papers as those of Lehman and Westoll (166) and Säve-Soderbergh (167). An amusing discovery was that of "fossil" lungfish burrows (Romer and Olson 1954) in the Texas Permian; since then, similar burrows have been recognized in other Permian and Carboniferous localities. Apparently the aestivating habits of dipnoans developed at an early stage.

Lower actinopterygians.—The ray-finned fishes of today are numerous and form an almost insoluble complex from the points of view of systematics and phylogeny. Although rare in the Devonian, the ancestral members of the group rapidly attained great prominence and variety, and their early history presents as many problems as does that of their modern representatives. Study of lower actinopterygians, a very active field in the prewar period, has continued to be an area of high interest. Gross

(1953) described Devonian palaeoniscids, and Lehman (126) described the skull of the oldest genus, *Cheirolepis*. Works primarily anatomical deal with the palaeoniscoid snout (Gardiner 132), *Bobasatrania* (Lehman 1956*a*), subholosteans (Schaeffer 125), *Birgeria* (Nielsen 1949), *Lepidotus* (Malzahn 1963), *Ichthyokentema* (Griffith and Patterson 141), Jurassic holosteans (Rayner 133), some Liassic forms (Gardiner 138), and jaw evolution (Schaeffer and Rosen 140); faunal papers in which lower actinopterygians are numerous include studies on the Triassic of Madagascar (Lehman 46, 47), on Mesozoic Congo faunas (Saint-Seine and Casier 51; Dartevelle and Casier 48), and on the Cerin lithographic limestone (Saint-Seine 50).

Before the fossil story came into prominence it was customary to divide the ray-finned fishes into three subgroups, Chondrostei, Holostei, and Teleostei. (These names are essentially meaningless today, since they implied that ossification was increasing as we ascended their evolutionary ladder.) The Chondrostei (apart from *Polypterus*, discussed later) include, of living forms, only the sturgeons and paddlefishes, primitive in, for example, the retention of a sharklike heterocercal tail and of a clavicle in the shoulder girdle, but degenerate (rather than primitive) in the feeble ossification of the internal skeleton. The Holostei include today only the gar pikes, *Lepisosteus* and relatives, and the "bowfin" *Amia*; these are fishes with more advanced characters such as the strong reduction of the heterocercal tail and loss of the clavicle. And, finally, there is the great "modern" group of the Teleostei, with the typical homocercal tail structure.

As fossil forms were discovered, it was but natural that attempts were made to fit them into this three-stage series. In the Paleozoic were found types—many at first included in the single genus *Palaeoniscus*—that differed from the living chondrosteans in better ossification and possession of thick shiny ganoid scales, but nevertheless in the fully heterocercal tail and other features were obviously primitive and could therefore be classed as chondrosteans. Later in the stratigraphic sequence—notably in Jurassic seas—were numerous forms which appeared to be about at the *Amia* or *Lepisosteus* evolutionary level, and hence could be placed in the Holostei. As time went on and discoveries increased, however, the situation became increasingly complex and confused. The ancient types with, in general, fully heterocercal tails and shiny ganoid scales increased continually in variety of known forms, so that from a single genus, *Palaeoniscus*, they became a family and today, with probably close to 200 genera, they rank at ordinal level or above.[2] Further, the fossil forms refused to sort out clearly into components of the three grades. In the Triassic, as Brough (134) first pointed out, there are various forms which were progressing structurally from the palaeoniscoid to the holostean level, although for the most part they were not holostean ancestors. To these Brough applied the term "Subholostei." In the later Mesozoic, again, there are numerous forms, such as *Pholidophorus*, *Leptolepis*, and *Lycoptera*, that appear to lie close to the holostean-teleostean boundary and are sometimes separated as a group of "halecostomes."

How to deal with the complex and put them into some meaningful systematic arrangement? One possibility is to follow Berg (32) and simply list a long series of orders, *seriatim*, without commitment as to their relationships or phylogeny. Any attempt at retention (or erection) of superordinal groupings here is subject to encountering the pitfalls of parallelisms. But I have, nevertheless, as an attempt toward retention of phylogenetic meaning, conserved the classical system and placed all actinopterygians in the three longstanding subgroups of Chondrostei, Holostei, and Teleostei. In the Chondrostei I have included, in addition to the living sturgeons and paddlefishes (and polypterids), all the "*Palaeoniscus*" group, using Palaeonisciformes in a broad ordinal sense. I have also placed here, as Redfieldoidei and Perleidoidei, most of the subholosteans. Of the forms I included in the Holostei, the subholostean Parasemionotoidei are generally believed to be ancestral to the typical *Amia*-like holosteans. Certainly on a holostean level are the Semionotidae, although it is quite probable that they have arisen from subholostean ancestors independently of the more typical holostean types. The deep-bodied coral-nibbling pycnodonts are of uncertain lineage; they are often thought to be

[2] Since I am not myself a worker in this area, I found it difficult to classify the ancient chondrosteans—palaeoniscoids in a broad sense—and to assign genera to families. As the present volume was about to go to press, there appeared a welcome paper by Gardiner (1967) with a reasonable classification of these forms, in which (except for a dozen or so forms, notably Rusconi's) nearly all genera are accorded a position in families.

derived directly from deep-bodied Paleozoic forms (most recently by Lehman 1956*a*; Gardiner 138), but alternatively they may be derived from semionotids, and justification for their position in the Holostei would depend on the place assigned the semionotids themselves. For the most part I have retained the "halecostomes" in the Holostei as a group of pre-teleosts.

Because of the fleshy fins and the presence of lungs, *Polypterus* and *Calamoichthys* were in early days considered to be crossopterygians. However, Goodrich (127) in 1927 showed very clearly that in basic features they are clearly actinopterygians, readily derivable from palaeoniscoids. The presence of lungs is presumably related to the possession of these structures by early bony fish generally; and although most actinopterygians show little development of a fleshy fin, White (130) has shown such a development in the palaeoniscoid *Cornuboniscus*. In consequence, it seems clear that *Polypterus* should be placed among the lower actinopterygians, and the placing of this form in a separate group of bony fishes, as a class or subclass Brachiopterygii, is an anachronism.

Teleosts.—Where to place the base of the Teleostei is problematical. It is generally agreed that the teleosts are derived from the pholidophoroids, which in turn are rather generally believed to have been derived (perhaps directly) from parasemionotids (Gardiner 138; Griffith and Patterson 141). Nybelin, who is currently studying forms in the holostean-teleostean transitional area (1961*a*, 1961*b*, 1963, 1964, 1966), believes that the Teleostei are to some extent polyphyletic in that several lines may be traced back independently to the Leptolepidae, now often included in the "halecostomes." The leptolepids have, in the past, frequently been considered as teleosts, and to remove much of the "threat" of polphylety, I have in my classification restored them to that group. I note, however, that Lund (1966) finds that intermuscular bones—generally considered a teleost character and present in *Leptolepis*—are already present in *Pholidophorus*.

Among works of most recent years concerning fossil teleosts may be noted Patterson's (146) important work on primitive acanthopterygians and a description of the apparently holostean-teleostean transitional form *Lycoptera* by Liu and others (1963).

Teleost otoliths appear to be useful for stratigraphic purposes, and a number of works have appeared on this subject. Of a modest number of broad faunal papers in the last two decades, we may note those by Arambourg (54), Arambourg and Signeux (53), Tabaste (1964), Casier (1961), and D'Erasmo (1946); extensive Tertiary European faunas are described by Leriche (57), Casier (55), Leonardi (1959), and Danil'chenko (1960).

When the last edition of my *Vertebrate Paleontology* was published, I found no classification of teleosts which seemed at all satisfactory, those available being in the main adaptations of those of Jordan or (rather better) Tate Regan, or that of Berg (32) who (as with lower fish groups) merely listed *seriatim* an unnecessarily large number of orders on the teleost level. Accordingly, I "rolled my own" and produced a classification that brought forth rather violent reactions from my friends among the ichthyologists.

Possibly such reactions may have been one of a number of factors which led to a project by the ichthyological members of the Society of Ichthyologists and Herpetologists to develop cooperatively a new classification. For a time this went well and several valuable preliminary reports were produced. Soon, however, the project ground to a halt. Much of the difficulty, I believe, lay in the fact that the group had, all too ambitiously, decided not to limit themselves to teleosts, but to reclassify all fishes. To properly evaluate the teleost situation would have been a major task; to include all fishes in a revision would have required the close cooperation of numerous paleontologists as well as students of living forms.

As a result of this failure, I found myself in the same situation in preparing a new edition of my book that I had encountered twenty years before. But, by a happy accident, I discovered, shortly before it became necessary to close my text, that a radically new teleost classification was in course of preparation by Greenwood of the British Museum, Rosen of the American Museum, New York, Weitzman of Washington, and Myers of Stanford. They were so kind as to give me access to their data in its then provisional form. Although they do not claim perfection for their work, it was obvious to me that, on a morphological basis, it was far superior to any previously in existence.

I therefore adopted essentially this system of classification for my new edition. Their work has now been published (142); apart from including the leptolepoids in the teleosts (as noted previously), the classification which I have used is almost exactly that proposed by Greenwood *et al.*[3]

Nearly all recent classifications of teleosts have, in "horizontal" fashion, tended to include all relatively primitive Recent forms and their presumed ancestors (notably a great array of Cretaceous fishes) in a basal order Isospondyli. The Greenwood group has, in contrast, attempted a vertical classification, and cleaves presumed primitive forms into three (or four) divisions—Elopomorpha (with which the Clupeomorpha are associated with some doubt), the Osteoglossomorpha, and a third division, of which the Protacanthopterygii are regarded as primitive. The authors are primarily concerned with living forms and have paid little attention to the fossils. I have attempted to sort out the more ancient "isospondyls" into groups antecedent to each of the three divisions, but do not feel too sure of my grounds, and very probably have erred badly. Of the three divisions, the first two are assumed to have had relatively little phylogenetic importance. The Elopomorpha have as a primitive "generalized" component the Elopoidei, including the modern tarpons and *Albula* and a number of related fossil forms. Grouped with them, surprisingly, are the eels (Anguilliformes)—an association based, despite the strong contrast in many regards, on the presence of a considerable series of common (and generally primitive) morphological features and the presence in these orders of a leaflike *Leptocephalus* larva. Appended, too, are the relatively unimportant Notacanthiformes. The Clupeomorpha, the herring-like fishes, are generally considered to be close to the general ancestral stock, and certainly were developed in Cretaceous times; our authors, however, consider them to be a sterile group—possibly a side branch of the Elopiformes and of independent origin.

The second of the three divisions, the Osteoglossomorpha, is a very small one, including for a certainty only a small number of freshwater forms, notably the osteoglossids, mormyrids, and

[3] Most useful to those who (like myself) are not intimately familiar with the almost infinite variety of modern teleosts is the inclusion in this publication of thumbnail figures of representatives of every Recent family.

gymnarchids of the tropics. They have a very distinctive series of primitive specialized characters, setting them off clearly from all other major groups. We cannot be sure of their ancestry. Possibly the marine Cretaceous ichthyodectids may be related; but this is very far from certain.

The greater part of all teleost assemblages are placed in Division III. The presumed ancestral group is the superorder Protacanthopterygii, in which the Salmoniformes (*sensu lato*) are emphasized as a truly primitive group rather than relegated (as if often done) to a secondary position, and the myctophids (scopeloids) are considered as occupying a central position with regard to the origin of advanced groups. The Ostariophysi are accorded superordinal rank; there are no living or fossil intermediates between them and other major groups, but it is pointed out that the gonorhynchiform order of the Protacanthopterygii show characters suggesting a structural ancestry.

Other forms below the acanthopterygian level are placed in two superorders—the Atherinomorpha, including only the order Atheriniformes (flying fishes, half beaks, cyprinodontids, etc. [Rosen 1964]), and the Paracanthopterygii. The latter, as the name suggests, includes a "mixed bag" of forms, such as the codfishes and anglers, which have approached closely the acanthopterygian level of organization in many regards but do not appear to be properly includable in that group phylogenetically, and have presumably arisen, in parallel fashion, from myctophoid salmoniforms.

The great group of spiny-rayed teleosts are here included in a superorder Acanthopterygii. It has long been suggested that the myctophoids are related to their ancestry; our authors, following C. Patterson (146), agree that the actual forms making the transition are not the myctophoids themselves, but the little derived group of Cretaceous Ctenothrissiformes. As is generally agreed, the Beryciformes are considered as a basal order. At this level, it is suggested, there branched off a number of primitive but aberrant groups, such as the Lampridiformes and Gasterosteiformes. The Scorpaeniformes are regarded as an advanced but specialized spiny-finned order. All other acanthopterygians are "lumped" in an enormous and highly complex order Perciformes. As Patterson (146) has emphasized, and as our authors agree, the Perciformes are probably polyphyletic in their origin from the primitive acanthopterygian stock, but as yet we do not have sufficient knowledge to sort them out.

As we have outlined it, the Greenwood-Rosen-Weitzman-Myers study gives us a currently workable classification of teleosts upon which future studies may well build for improvement and modification. It is, however, almost entirely based on living forms, and correlated studies of the earlier fossil teleosts are much to be desired.

Are the teleosts polyphyletic? If the leptolepids and other forms of a comparable primitive nature be excluded from the Teleostei, the group probably is, as our authors suggest. But there is no clear indication of origin from two or more distantly related groups on the holostean evolutionary level, and as far as present evidence goes, there is no indication of any strong polyphyletism in the group.

Most of our Mesozoic materials of teleosts come from marine deposits; most major teleost groups (with the notable exception of the Ostariophysi) are exclusively or nearly exclusively marine in modern distribution. We hence tend to get a strong impression that teleost evolution was essentially a saltwater phenomenon. Our outlook tends to be strongly biased by the paucity of freshwater fish-bearing beds of late Mesozoic age, however. If and when major deposits of this sort should be discovered, the picture may change, and we may find that much of the evolution of teleosts, including groups other than the Ostariophysi, took place in inland waters.

6

Amphibians

During the past two decades there has been a great amount of work on fossil amphibians, with discovery of much new material, restudy of old materials with more modern methods, and considerable theoretical discussion as to the interpretation of the record. In this period there have been several broad reviews of the subject, including comprehensive accounts in Huene's volume on lower tetrapods (226), an extensive review by Piveteau and Lehman in the Piveteau Traité (1), and a survey of much of the field in my review of labyrinthodonts (173) and discussion of microsaurs (209).

Before discussing specific evolutionary problems, we should mention two stumbling blocks—one as to materials, one mental—in our attempts to unravel the complexities of amphibian phylogeny. It seems certain that nearly all major phyletic lines among the lower tetrapods were established before the close of the Carboniferous, and hence adequate knowledge of the life of that period is vitally important. We are greatly hampered in such studies because known Carboniferous fossil material is limited and, further, because much of the known material has not been, as yet, adequately studied by modern methods. Almost without exception, known tetrapod fossil localities of that age are located in western Europe and eastern North America; we know nothing of Carboniferous or evolutionary events in the rest of the world. We are limited as to time as well as space; almost all these sites lie stratigraphically in the upper part of the Carboniferous sequence, and described material from the Lower Carboniferous is as yet limited to a few specimens from Scotland. (Some Mississippian material from West Virginia awaits description.) The materials come mainly from (1) Nýrany, Bohemia; (2) a number of localities in Great Britain; (3) Linton, Ohio; (4) the erect trees of Nova Scotia. Panchen and

Walker (1960) have published a valuable paper giving the actual stratigraphic position of British labyrinthodont occurrences, correcting common mistakes regarding various localities. In early days, Fritsch gave an extended account of the Bohemian finds; British specimens were mainly described by Huxley; Cope published on the Linton fauna; and Dawson on Joggins tree stump materials. Later, in the 1920's and 1930's, Watson published valuable redescriptions of British specimens (Watson 199, 200); Steen did useful work on Bohemian and American amphibians (Steen 1931, 1934, 213); I contributed to some extent on revision of the Linton fauna (Romer 1930), in which, although correcting a number of older mistakes made by Moodie (1916), I committed some new ones myself. All these materials need, however, after the lapse of several decades, careful reworking with improved methods and from a modern point of view. As this is written such work is under way by a number of workers—for example, by James and Margaret (Steen) Brough on certain Scottish and Bohemian forms, by Alec Panchen on British labyrinthodonts, by Robert L. Carroll on Nova Scotia specimens, and by Donald Baird on Linton amphibians. Future publications by them will, we may be sure, add to our understanding of the story. But we desperately need new discoveries which will broaden our base in space and time.

Vertebral structure looms large in any study of ancient amphibians. A mental handicap has been placed on our understanding of early tetrapod history because of general adherence in the past to Gadow's thesis that the vertebral column in all vertebrates is constructed according to various permutations of a basic pattern, in each segment, of four pairs of arcualia: basi- and interdorsals in the arch region, basi- and interventrals in the region of the centrum (Gadow 1933). Workers on the embryology of the column in living tetrapods have attempted, without conspicuous success, to interpret their results in terms of this theory; and still greater confusion has been made by attempts in paleontology to interpret the vertebrae of fossil forms in terms of Gadow's four pairs of hypothetical arcualia. It now seems clear that (whatever may be true of fishes) there is no evidence in the embryology of living tetrapods to support the theory of four pairs of arcualia (Williams 175). To be sure, there is in all higher vertebrates an interesting cleavage of the sclerotomic tissue of each segment and a pairing of each of the halves so formed with one derived from a neighboring segment; but this process shows no relation to Gadow's supposed arcualia. Nor does the fossil evidence furnish any support. At the most there may be three elements, or pairs of elements, in a vertebral segment: dorsally (1) the neural arch; ventrally (2) a single intercentrum; and (3) pleurocentra, primitively paired, from which, phylogenetically, the amniote centrum develops. Further, a fraction of our oldest amphibians show but a single ventral element, a spool-shaped "holocentrum"; various attempts to identify this structure with either of Gadow's two theoretical ventral arch pairs are valueless.

Tetrapod origins.—In the last century it was commonly believed that tetrapods arose from the dipnoan fishes. Numerous features in the soft anatomy of lungfishes and amphibians are very similar, and the embryological development of the two groups is similar indeed. But dipnoans are highly specialized in cranial and dental structures, and early in the present century the work of Goodrich, Watson, and Gregory, among others, made it apparent that we must look for tetrapod ancestors among the Crossopterygii.

In view, however, of the fact that parallelism is known to have been a common feature in evolutionary history, it is not surprising that the possibility of dual origin of tetrapods has been suggested. Some decades ago Wintrebert (1922) and Holmgren (1933), followed by Säve-Söderbergh (1934), suggested that while frogs and all higher vertebrates had arisen from crossopterygians, the urodeles were dipnoan descendants. However, this theory was never fully elaborated, and failed of support. Wintrebert, essentially, merely argued that since the palate in the two modern groups differed considerably, they might have come from two different fish groups. Holmgren noted differences in tarsal construction and argued that while the frog limb might have come from crossopterygians with an abbreviated archipterygium, the urodele ancestor must have been a form with a fully developed archipterygium, such as was present in the older Dipnoi. But as Schmalhausen (1959) has pointed out, there is little strength in Holmgren's argument;

further, typical archipterygia are present in some crossopterygians as well as lungfish. No one, I think, would currently advocate the descent of any tetrapod from dipnoans.

Jarvik (150), however, came forward with an alternative theory of the dual origin of tetrapods that advocates such origins from two branches of the rhipidistian stock. He engaged in the detailed study of the snout region of *Porolepis*, an early Devonian relative of *Holoptychus*, contrasted his findings with what was known of other crossopterygians (mainly *Eusthenopteron*), and, as noted earlier, set forth the thesis that the rhipidistian crossopterygians could be divided into two groups, which he termed the Porolepiformes and Osteolepiformes. The two he believed to contrast strongly in a variety of features, notably the presence or absence of a precerebral area of the endocranial cavity and the presence of two types of pockets in the roof of the mouth which were assumed to lodge glandular structures. He further claimed that the structures seen in the snout of *Porolepis* showed similarities to those of urodeles, while those in the osteolepiform snout were anuran-like. He therefore concluded that the two amphibian types had originated from two different crossopterygian stocks. Jarvik has continued his work on the anatomy of *Porolepis* and its relatives with interesting and valuable morphological results (153, 154). But studies of crossopterygian structure by other workers (Kulczycki 155; Vorobjeva 1960; Thomson 156, 157, etc.; Schaeffer 171) have failed to substantiate his conclusions about tetrapod origins. For example, the variable pits in the mouth roof of rhipidistians, which he compared with those lodging mouth glands in amphibians, appear to be, actually, for the reception of lower jaw tusks; the presence or absence of a precerebral cavity is not diagnostic of any major group.

Jarvik (1966) has replied with vigor to his critics, particularly criticizing Thomson's work on *Ectosteorhachis* through his own work on *Megalichthys*, which he claims is generically identical (the two are not identical, although they are related). He admits (1966, pp. 86–90) that the snout pockets are primarily for tusk reception rather than for glands, and that the supposed group difference in the extent of the cranial cavity does not hold (p. 53, footnote 2). Thus the most prominent points of supposed difference are nonexistent. Although the early appearance in the Carboniferous of lepospondylous amphibians as well as labyrinthodonts leaves us with major unsolved problems regarding tetrapod early history, there is no reason to believe in a dual origin for tetrapods of the type envisaged by Jarvik. Further, basic in his argument is the assumption that urodeles and anurans are widely separated stocks. But the work of Parsons and Williams, discussed below, suggests that they may actually be related. If so, the whole question of a "dual origin" is pointless.

Although there is now essentially universal agreement that the tetrapods arose from rhipidistian crossopterygians, the structural gap between fish ancestors and even the most primitive of known amphibian descendants appeared great to workers in the early decades of the century. Seemingly a major stumbling block to derivation of amphibians from crossopterygians lay in the peculiar nature of the braincase in that fish group. It consists of two distinct moities, anterior and posterior, the two somewhat moveable on one another, whereas all other proper vertebrates have a solid braincase structure. In further contrast to all other forms, the posterior segment of the crossopterygian braincase is traversed ventrally by a large tunnel occupied by an anterior extension of the notochord. Could the normal type of braincase structure present in all tetrapods have possibly been derived from such a specialized type? As noted below, the discovery of the skull of ichthyostegids has in great measure solved this problem.

Some decades ago there appeared to be a major difficulty in comparing the skull roofs of rhipidistians and crossopterygians. In tetrapods (although there are a few variants) the opening for the pineal eye lies well back on the skull between the two parietal bones. In rhipidistians, in seemingly strong contrast, it lies far forward, between a pair of bones which were customarily called the frontals. How could this difficulty be resolved? Säve-Söderbergh (1934) suggested that both frontals and parietals were compound elements, each with anterior and posterior components, and that the apparent differences between rhipidistians and amphibians in the position of the pineal opening were due to differences in the way in which these hypothetical components had fused in the two cases. Little evidence was furnished to support his assumption, however, and his effort to resolve the difficulty, essentially by the use of shears and paste-pot, met with little acceptance. The clue to the solution was furnished by me in a paper on the dipnoan skull (1936) in which, comparing lungfish and crossopterygian roof patterns, I pointed out that the "extra-scapular" bones of crossopterygians, usually considered homologous with the tabulars and postparietals of tetrapods, were not properly part of the skull roof but were merely enlarged anterior scales. Hence it was the bones next anterior to these elements—the ones usually called parietals—that were actually the homologues of the tetrapod postparietals. Stupidly, with my attention at the time on dipnoans, I did not realize the implications of this. Westoll (1938) did, however, and demonstrated that the shift in the position of the pineal from fish to tetrapod was not due to a shift from one pair of bones to another, but to a shift in the proportions of the skull regions and the associated dermal bones. In the transition from crossopterygians to typical amphibians, the snout region is relatively elongated, the postorbital region has become relatively short, and the postparietals, originally very large and long elements, have been much reduced. The pineal-bearing bones have moved backward in relative position. The so-called frontals of crossopterygans are actually parietals. Once Westoll had shown this, I (1941) was able to carry the picture of change onward into higher tetrapods, where further reduction of the postparietals has occurred, and the parietals, once relatively far forward, occupy the very back end of the skull roof.

Although the true interpretation seems perfectly evident Jarvik (8) has clung to the interpretation of the fish pineal-bearing bones as the frontals. His evidence, however, is exceedingly slight (cf. Romer 1962*a*).

The oldest tetrapods.—The diversity of Carboniferous amphibian types indicated that the evolution of land forms must have begun well back in the Devonian, but it was not until 1932 that Säve-Söderbergh reported the discovery of amphibian skulls from East Greenland beds close to the Devonian-Carboniferous boundary. The nature of the specimens then known permitted little but a description of the skull roof of *Ichthyostega* and a related genus. Since the last war, much new material has been found, permitting description of the entire skeleton; so far, only preliminary descriptions have been given (Jarvik 177, 178). Postcranially, the ichthyostegids are already well over the fish-tetrapod boundary; a vestige of fish median fin persists, but the limbs, albeit very short, if stocky, are comparable to those of the more primitive amphibians of later times. The skull is most interesting, for the braincase structure is transitional between that of rhipidistians and of later labyrinthodonts. The two braincase halves appear to have united, but a distinct line of suture between them persists; however, here, in contrast to all other known amphibians, the rhipidistian notochordal canal is still present! The skull roof pattern, particularly in the absence of an intertemporal bone, suggests that the ichthyostegids are slightly off the main line leading to later labyrinthodonts.

The fact that the ichthyostegids had already achieved the truly amphibian level in limb and girdle structure makes it clear that tetrapod development must have begun at a much earlier stage of the Devonian. But, to date, the only bit of evidence of this is a single specimen showing a partial skull roof suggestive of a rhipidistian-labyrinthodont transitional condition, from the earlier Upper Devonian horizon of "Scaumenac Bay" (Westoll 1938, 176).

Soon after Säve-Söderbergh published his descriptions of the first finds of ichthyostegids, it became apparent to me that *Otocratia*, described by Watson (200) on the basis of a Lower Carboniferous skull from Scotland, was a close relative, and Westoll (1942*b*) reached the same conclusion. The thesis of ichthyostegid relationship appears to be confirmed. *Otocratia* shows an interesting specialization, described by Watson (200), of the tabular and otic notch region. Among the new materials of ichthyostegids described by Jarvik is *Acanthostega*, represented (as far as described) only by the skull roof. It is incomplete posteriorly, but appears to show a peculiar condition which (although this is denied by Jarvik [177]) appears to be basically similar to the *Otocratia* specialization here.

Major labyrinthodont classification.—It is a curious fact that despite gradually increasing knowledge, the story of amphibian evolution is still divisible into three discrete compartments: (1) the story of the Labyrinthodontia, most prominent of Paleozoic and Triassic amphibians, which had an interesting history of their own, show links with crossopterygian fishes, on the one hand, and with reptiles, on the other, but remain without known connections with other amphibians of any age; (2) the story of the Paleozoic lepospondyl orders, which flourished in modest fashion in the Carboniferous and early Permian, but show no evidence of connections with fish ancestors or with contemporary labyrinthodonts, and no clear indications of relationships to later forms; and (3) the story of the modern amphibian orders, of which the anurans and urodeles can be traced back into the Mesozoic, but without giving indication of their earlier pedigree.

In earlier works (173, etc.) I generally included the labyrinthodonts and anurans in a subclass, Apsidospondyli, on the assumption that the frogs showed in certain cases at least traces of the presence of central "arch" elements somewhat comparable to the situation in labyrinthodonts, implying a special relationship of anurans (in contrast to urodeles) to the former group. Currently there appears to be no basis for this assumption, and I have therefore abandoned the term Apsidospondyli.

The labyrinthodonts, highly varied in form and structure, and abundant in numerous fossil deposits of the late Paleozoic and Triassic, have received much attention from paleontologists for more than a century. Ideas concerning their relationships and evolution remained, however, in a most chaotic condition until Watson in his classic papers of 1919 and 1926 (180, 199) pointed out that in general three evolutionary grades could be made out, in a sequence from the Carboniferous through the Permian to the Triassic. Carboniferous forms were, in general, primitive water dwellers, with relatively high skulls, a single occipital condyle, and an essentially closed palate moveably articulated with the base of the braincase. In the Permian some (but far from all) labyrinthodonts had moved a long way toward terrestrial life, but on the whole the skulls had become rather flattened, the occipital condyle was tending to divide in two, and the palate was moderately fenestrated and fused, although often narrowly, with the braincase. In the Triassic, the remaining labyrinthodonts had slumped back to a purely water-dwelling life, the skull was greatly flattened, the condyle was double, there were very large interpterygoid vacuities, and the palate was broadly welded to the braincase floor.

Although with qualifications and exceptions, this general picture of labyrinthodont evolution remains as acceptable as when first set forth by Watson several decades ago (cf. also Watson 13, chaps. 2–4). One further thesis proposed by Watson, however, has not been borne out by later work; this was the attempt to correlate the evolutionary stages with vertebral types. Three types are commonly found in labyrinthodonts: (1) rhachitomous—the centrum is mainly composed of a large intercentrum, forming a ring which is incomplete above, but includes also a pair of small pleurocentra, situated posteriorly and dorsally; (2) stereospondylous—the pleurocentra have disappeared and in the central region an intercentrum only remains, sometimes expanded dorsally to form a complete ring; (3) embolomerous—both intercentrum and pleurocentra form complete rings. The rhachitomous type is prevalent in the Permian, and the stereospondylous type typical of the Triassic. At the time Watson wrote, the only Carboniferous labyrinthodonts in which the vertebrae were well known showed an embolomerous structure. He therefore concluded that the Embolomeri were the truly primitive labyrinthodonts and that the sequence ran: Embolomeri—Rhachitomi—Stereospondyli, with the reptiles (with a reduced intercentrum and a true centrum of pleurocentral origin) branching from the embolomeres.

Watson's demonstration of the series of structural stages was so well documented and convincing that all workers tended to accept, in addition, his beliefs about the evolution of vertebral structure. As time passed, however, it became more and more difficult to hold to his assumption that all Carboniferous labyrinthodonts were embolomeres and that the Embolomeri were the basal labyrinthodont stock. As our knowledge of Carboniferous forms increased, there were found numerous instances in which forms of that age showed cranial patterns of rhachitomous type, and certain, at least, of these were also rhachitomous, rather than embolomerous, in vertebral structure. As a result I proposed in 1947 (173) a new theory of the evolution of vertebral types amongst the Labyrinthodontia: that the primitive amphibians had not embolomerous, but rhachitomous vertebrae (already foreshadowed among the rhipidistians) and that the embolomeres, although primitive in many regards, were not the ancestral amphibians but pertained to a branch leading toward the reptiles. This theory proposed that (apart, perhaps, from the ichthyostegalians) all labyrinthodonts be ranged in two major groups, Temnospondyli and Anthracosauria. The temnospondyls, forming the "main line" of labyrinthodont evolution, began with Carboniferous types of primitive structure but with rhachitomous vertebrae, evolved in the later Carboniferous and Permian into typical rhachitomes, and eventually gave rise to stereospondylous forms. In this line the intercentra, large to begin with, remained the dominant element in centrum construction, and the pair of little pleurocentra tended to disappear in later stages. The Anthracosauria were fewer in numbers and had a shorter career, but were phylogenetically important for including not only the embolomeres as a side branch but the line leading to reptiles. Here, in contrast to the temnospondyls, the intercentrum tends toward reduction (except in embolomeres) while the little pleurocentra grow and fuse to form the main centrum element.

Since this thesis was put forward, numerous discoveries have tended to prove its validity. For example, the vertebral column of the ichthyostegids (unknown at that time) proves, as this theory demanded, to have a protorhachitomous structure; as noted below, the Carboniferous loxommids, which Watson believed to be embolomeres but whose skulls show the pattern expected in a primitive rhachitome, are now known to have rhachitomous vertebrae, as do a number of other Carboniferous labyrinthodonts; we are beginning to find among fossil forms the structural stages expected by this theory in the development of anthracosaurian vertebrae.

Temnospondyl history.—The temnospondyl group includes the vastly larger fraction of the labyrinthodont assemblage; a large part of the known Carboniferous forms are now seen to be temnospondyls, as are all but a dozen or so Permian genera and all those of the Triassic. Their long evolutionary history was obviously a complex one. Year by year new discoveries are being made, and their phylogenetic pattern is gradually emerging. Watson long ago pointed out the existence of a central evolutionary line leading through such typical rhachitomes as *Eryops* of the early Permian on towards such advanced Triassic stereospondyls as the capitosaurs. There were, however, numerous side branches and variants. Watson's belief as to the evolutionary pattern here can be crudely compared to a bush with long vertical shoots along which various lines evolved in parallel fashion; for example, he believes that most of the numerous short-skulled forms, known from the Carboniferous to the end of the Triassic, were members of a single series paralleling the "main line" (Watson 182). My own current concepts suggest a more treelike pattern, with side branches emerging from the main stem at successive levels. I suspect that the true picture may eventually prove to be one somewhat intermediate between our points of view.

The loxommids, readily identified by their keyhole-shaped orbits, include (except for *Pholidogaster*) the oldest-known Carboniferous labyrinthodonts of which we have any adequate knowledge. Since they were Carboniferous in age, Watson assumed them to be embolomeres; nothing was known about their postcranial structure.[1] But as I pointed out in 1947 (173) the loxommids, apart from enlargement of the orbit, exhibit in every character, as the closed palate and so forth, exactly the structure which one would expect in a basal member of the temnospondyls. Such animals should have been common in the Lower Carboniferous, but knowledge of them is denied us because of a dearth of material of that age. That my deduction was correct is proved by the discovery by Baird (1957) of rhachitomous vertebrae in an American loxommid.

In a classification of amphibians made in 1960 (p. 20), O. Kuhn makes the curious statement that since the loxommids have been shown not to have embolomerous vertebrae, the term Embolomeri should be abandoned! This is a *non sequitur*; the term Embolomeri was based by Cope on *Cricotus* (and *Archeria*), whose vertebrae are just as embolomerous as ever; Watson included the loxommids, hypothetically, among the embolomeres on purely stratigraphic grounds.

Ascending from the archaic, loxommid, level of temnospondyl evolution, we reach that of the characteristic Rhachi-

[1] In 1884, Baily described (but unfortunately did not figure) the skeleton of a large amphibian from Kilkenny which, as well as one can interpret his account, would appear to have been a loxommid. Search in Dublin, however, failed to uncover any trace of this specimen; rumor has it that it was used for fuel during a hard winter at the time of the troubles in the early 1920's.

tomi. Watson, both in his earlier papers and in a recent work (181) has ably outlined the general progress of the temnospondyl "main line" upward through the rhachitome and stereospondyl evolutionary levels. In 1947 (173) I attempted to establish three ascending levels among the Rhachitomi proper—edopoid, micropholoid, eryopoid. In the edopoids the palatal vacuities are of modest size, the basal articulation is moveable, and the archaic intertemporal bone persists. In the eryopoids there is a modest degree of enlargement of the interpterygoid vacuities, palate and braincase are immoveably joined (although the connection is a narrow one), and the intertemporal is lost. The micropholoids were assumed to be transitional between the two, with the intertemporal bone lost but the moveable basal articulation retained.

Recent work indicates that my supposed micropholoid group is an artificial assemblage of odds and ends (Carroll 179; Langston 187); the fusion of palate and braincase may occur ontogenetically or be lost neotenously. Successive edopoid and eryopoid stages appear, however, to be valid. The edopoids are best known, cranially, from *Edops*, which occurs at the very base of the Permian but is essentially a last survivor of a Pennsylvanian type of relatively primitive rhachitome. *Dendrerpeton*, of the early Pennsylvanian of the Nova Scotia coal measures (studied by Steen in 1934 and now being restudied by Carroll), is not improbably a central form of this group. Steen restored for *Dendrerpeton* a vertebral structure of uncommon type; her material, however, was not articulated, and probably the elements she restored as alternating intercentra and pleurocentra, both large ventral crescents, are merely intercentra, small pleurocentra presumably having been present in life but not identified and perhaps unossified. (Since writing the above, an excellent redescription of *Dendrerpeton* by Carroll [1967*a*], confirms this conclusion.)

Several side branches appear to have sprung from the rhachitome stem at this level. Prominent are members of the *Trimerorhachis* group, primitive in certain regards but short-faced (although long postorbitally) and with a strong trend toward skull flattening and the development of large palatal vacuities. Trimerorhachids were most abundant in the early Permian, but *Saurerpeton* (*Branchiosauravus*) of the Linton cannel is a Pennsylvanian forerunner. Watson (182), basing his work primarily on the early Permian trimerorhachid *Eobrachyops*, suggests (following his general tendency toward belief in "bush" rather than "tree" evolution in the temnospondyls) that all later short-faced forms stem from these early types. *Dvinosaurus*, a neotenous form from the Upper Permian of Russia, may be such a descendant. But it is a far cry from a long-skulled trimerorhachid to a short-skulled brachyopid, and quite surely, as discussed later, the plagiosaurs are members of a different evolutionary line. Of Triassic temnospondyls, the only ones with trimerorhachoid skull proportions are the metoposaurs; but their ancestry is a blank.

Colosteus and *Erpetosaurus* are forms from the Linton cannel which I (1930) once suggested to be ichthyostegid descendants that had paralleled the temnospondyls in the development of palatal vacuities. This suggestion has met with little favor and, I expect, should best be abandoned. A redescription of these forms by Baird and Chase will be published shortly, I hope. In general, they greatly resemble the trimerorhachids, but the intertemporal is absent. In this regard they approach the metoposaurs more closely than the trimerorhachids; but the jump from Carboniferous to Upper Triassic without known intermediates is a long one.

Also diverging, it would seem, at the edopsoid level is *Chenoprosopus*, as redescribed by Langston (187). Several Carboniferous genera—*Cochleosaurus* (*Nyrania*), *Gaudrya*, and *Macrerpeton* (*Capetus*, *Mytaras*)—may be allied with it.

Eryops, the type eryopoid, is an early Permian form, and eryopoids are widespread in the early Permian; but a number of eryopoid genera have now been recognized in late phases of the Pennsylvanian. The group was, for a time, very successful, and in addition to a number of early Permian genera closely related to *Eryops*, there appears to have been a considerable radiation from the base of the eryopoid stock. Most notable are the dissorophids, recently studied by Carroll (179). A considerable variety of these forms developed in the Permian, and a Pennsylvanian forerunner was *Amphibamus*, apparently widespread and common, although in the past disguised nomenclatorially under a full dozen generic names. Interesting in the Permian is the trend for the development of armor in a number of genera, seemingly correlated with defense in these most terrestrial of labyrinthodonts. The peculiar long-spined *Platyhystrix* was at one time thought to be identical or close to *Zatrachys*, but Langston (187) has shown that the latter genus has a normal spinal column. *Zatrachys*, with its peculiarly spiny skull, well described by Langston, is a member of one of several other families radiating from the eryopoid stock. From the eryopoid level, or one slightly antecedent, appear to have sprung long-snouted fish-eaters of the *Archegosaurus-Platyops* group. Whether the similarly long-snouted trematosaurs, widespread in the early Triassic, sprang from this line or from a point higher in the temnospondyl sequence appears to be uncertain. *Intasuchus* and *Syndiosuchus* of the Permian (Konjukova 1953), despite a slight contact (as figured) of tabular and parietal, appear to be in other respects typical if rather long-skulled temnospondyls on an eryopoid level.

"Branchiosaurs."—Well back in the last century there began to be discovered in the late Pennsylvanian and early Permian of Bohemia, Germany, and France tiny amphibians, resembling miniature rhachitomous labyrinthodonts in many regards, but with poor ossification of the postcranial skeleton, very short broad heads, and, in many cases, remains of external gills comparable to those of modern larval urodeles. Because of this last feature they were reasonably termed "branchiosaurs." Credner, the major early worker on these forms, believed that they had semicylindrical ventral central elements of an unusual nature, and, based on this supposed distinctive feature, there was erected for the branchiosaurs the order Phyllospondyli. Except somewhat doubtfully by Thevenin, no later worker has reported vertebrae of this unusual structure in branchiosaurs, and in general there appeared to be no ossification of any sort in the region of the centrum.

Nevertheless, belief in the existence of a distinct order Phyllospondyli continued. Impetus was given the group by Watson's (1921) description of *Eugyrinus*, a small amphibian from the English Carboniferous with a palate like that of a rhachitome. According to Watson's interpretation of labyrinthodont history, however, rhachitomes should not have been present in the Carboniferous. The occurrence of such an animal was accounted for by his assignment of *Eugyrinus* to the Phyllospondyli, in which a palate of this type was present.

A great increase in the number of supposed Phyllospondyli was caused by reinvestigation of the important Carboniferous fauna of Linton, Ohio, by Steen (1931) and myself in 1930. We found a considerable number of forms with skulls like those of rhachitomes; since, under the then prevalent theory, rhachitomes should not be present in the Carboniferous, and since in the forms we were considering the vertebrae were unknown or poorly known, we "solved" the problem of their presence by including them in the ranks of the Phyllospondyli. The ichthyostegids, obviously not being embolomeres, although ancient, were thrown in for good measure. The order Phyllospondyli thus tended to become an exceedingly bloated group, with a host of other animals, large and small, in addition to the little branchiosaurs which were the original members of the group.

Later in the 1930's a typical gilled branchiosaur was found for the first time in American deposits (Romer 188). Study of this specimen led me to consider what the true nature of the "branchiosaurs" might be. Presumably the large labyrinthodonts began life as little gilled larvae, which would have looked much like a branchiosaur. But if, under proper conditions, the fossil record preserves the remains of small and delicate branchiosaurs, why do we not find larval labyrinthodonts as well? From this last query arose the final and crucial question: *are not "branchiosaurs" simply larval labyrinthodonts?*

A broad survey of our knowledge of Permian and Carboniferous amphibian remains led inevitably to the conclusion that this is the case, and that the order Phyllospondyli is a purely imaginary one. With growing recognition of the presence of rhachitomes in the Carboniferous, the accretions made by Miss Steen and myself to the phyllospondyls could be painlessly removed and assigned to the Rhachitomi. As for the core of the supposed order, the typical branchiosaurs, they can be reasonably considered as larval and growth stages of contemporary rhachitomes. The supposedly diagnostic features of the branchiosaurs included the short broad skull, with large otic notches, the absence, in general, of discrete central ossifications in the vertebrae, the presence of external gills, and the absence of an ectopterygoid. All these features had long been known in small and immature forms from Lebach, which have been assigned without question to the rhachitomes *Archegosaurus* and *Actinodon*. Surely the German and Bohemian materials described as branchiosaurs are also young of contemporary rhachitomes, such as *Onchiodon*, *Sclerocephalus*, and *Chelydosaurus*. Not all "branchiosaur" larvae need, of course, to have pertained to rhachitomes. The little "branchiosaurs" from Moravia most recently described by Špinar (206) are immature seymouriamorphans, some showing gills.

Despite the fact that the situation seems quite clear, the "Phyllospondyli" were included (although apparently with some hesitation) by Lehman in the Piveteau Traité in 1955. There were several reasons. One was that at that time Watson, whose opinion was highly worthy of respect, had not yet come to my point of view. But he has since done so (Watson 189). Further reasons were that Heyler (1957), in his then current studies of Autun "branchiosaurs," found a ventral ossification in the centrum of one vertebra of one specimen, and that the amount of variation existing among the numerous Autun specimens strongly indicates that they could not all be larvae of a single form. These arguments, however, are of no weight. Since intercentra must ossify during late larval stages of the development of a rhachitome, it is not surprising that Heyler found one instance of ossification; rather, it is surprising that he did not find more. Further, one would expect a variety of larvae in deposits of Autunian age, since a wide variety of rhachitomes was in existence at that time.

It is hard to overcome inertia, and it has been so long customary to include the Phyllospondyli amongst amphibian fossil groups that one hesitates to omit them. It is, however, high time that this imaginary order disappeared from texts and treatises.

Stereospondyls.—The type stereospondyls are the capitosaur-like forms of the Triassic, with flat skulls, double condyles, very large interpterygoid vacuities, a broad connection between palate and braincase, and (in theory) stereospondylous vertebrae in which the intercentrum has become a complete disc. But this last, and supposedly diagnostic, feature is now seen to be relatively rare; further, a whole series of forms, frequently termed "neorhachitomes," have been described from the later Permian and early Triassic, which, leading onward from eryopoids, approach the capitosaur level (cf., for example, Watson 181). Despite the fact that the pleurocentra still persist in many of these forms, I thought it reasonable in 1947 (173) to expand the term Stereospondyli to include the neorhachitomes as the superfamily Rhinesuchoidea, leading to the capitosaurs and their immediate ancestors (the Capitosauroidea). In recent decades our knowledge of capitosaurs has increased considerably, revealing numerous variations in the trend toward a closed otic notch; a recent paper of note is Watson's description of an Australian capitosaurid skeleton (193).

Of non-capitosauroids customarily included in the Stereospondyli, the metoposaurs occupy an isolated position; in their general organization they are (apart from orbital position) very similar to the capitosaurids, but there is no intergradation in known forms, and the metoposaur "advances" may have been made in parallel fashion. Sawin (195) has restored the postcranial skeleton of "*Buettneria*"; Colbert and Imbrie (196) have recently reviewed the American metoposaurs, with a welcome reduction of names to the single genus *Eupelor*; Chowdhury (1965) points out that even this may well be indistinguishable from the first-described genus *Metoposaurus*. (I have used this last name in my text, but may point out, for the benefit of those loving priority, that *Eupelor* Cope 1868 far antedates *Metoposaurus* Lydekker 1890; however, *Hyperokynodon* Plieninger 1852 may be a still older applicable name.)

Generally included in the Stereospondyli are the short-skulled brachyopids, mainly known from the Triassic of southern continents. In my review of labyrinthodonts I suggested that they be bracketed with the metoposaurs, both families being short-faced, although differing radically in postorbital development, but there is no evidence to support this; Watson, on the contrary, would align them with a long antecedent series of short-faced forms dating back to the Permian. Rusconi (1951) has described from the Argentinian Triassic a suite of forms with skulls of brachyopid proportions. Estes, however, who with Welles of Berkeley is currently describing Triassic brachyopid material from Arizona, has pointed out to me that Rusconi's specimens, although short-faced, appear not to be true brachyopid. Rusconi associated with his amphibians a postcranial skeleton which is definitely that of a thecodont reptile. Postcranial materials properly associated with specimens of Rusconi's "brachyopids" in my possession, as well as material in the Instituto Lillo of Tucuman, show a "neorhachitomous" vertebral structure, and Estes informs me that neorhachitomous vertebrae appear to be associated with brachyopid skulls from the Moenkopi now being studied by him and Welles.

Long associated with the brachyopids, but surely, on present knowledge, far removed from them phylogenetically, are the extremely short-faced plagiosaurs characteristic of the late Triassic. To Nilsson (198) is due our only adequate knowledge of the postcranial anatomy of these extremely odd creatures. The vertebrae are, so to speak, super-stereospondylous, the centra being rather elongate cylinders. Obviously these peculiar structures cannot be directly derived from the much more primitive centra of the brachyopids. Probable light on their origin is shed by Panchen's description (197) of *Peltobatrachus* from the Upper Permian of East Africa. In most regards this animal is closely comparable to an eryopoid such as the dissorophids, and a further suggestion of dissorophid connections is the fact that *Peltobatrachus* is armored (the ribs appear to be specialized in this connection). The centra, however, are not at all eryopoid and, like those of the plagiosaurs, are moderately long cylinders. It is not unreasonable to believe that, as Panchen suggests, we are dealing with a plagiosaur ancestor. As to the origin of this phyletic line, it is not impossible that this vertebral type has been derived from a dissorophid rhachitome of the late Carboniferous or even earliest Permian. Panchen advocates erection of a separate order for the plagiosaurs; however, despite the extreme specialization of the centra, the general organization of *Peltobatrachus* is so characteristically temnospondyl that a subordinal separation seems sufficient.

Panchen (1963) argues that the plagiosaur centrum is a pleurocentrum. I fail to see the force of his argument. Despite the fact that we know nothing of plagiosaur ontogeny, it is obvious that the centrum includes in its composition the embryonic materials which in other labyrinthodonts are divided between intercentrum and pleurocentra; if either of the two grows at the expense of the other, it gradually takes over a larger and larger portion of the total embryonic tissue. In a reptile it is the pleurocentrum which has done this; if, as seems reasonable, the plagiosaurs are derived from rhachitomous ancestors, it is more probable that the intercentrum has triumphed here—as is the case with the stereospondyls. The rib articulation also suggests this interpretation.

The Permian *Tungussogyrinus* (Efremov 1939) is perhaps a larval early plagiosaur. Alternatively, it may be a larval brachyopid.

Anthracosaurs.—In contrast to the temnospondyls (highly successful for the time, but doomed to eventual sterility) the anthracosaurs existed for a much briefer period and are represented by only a small number of forms. But, since they gave rise to reptiles, they are of great importance in the study of vertebrate phylogeny and anatomy. An unimportant but exceedingly useful and readily observed feature by which anthracosaurians can be distinguished from temnospondyls is that, as first noted by Säve-Söderbergh and Steen, the tabular of anthracosaurians has a contact, generally absent in temnospondyls, with the parietals. The key character of the group, however, lies in the development, from its beginning as a pair of small pleurocentra, of a stout true centrum. The intercentrum, in contrast with that of temnospondyls, is reduced in advanced anthracosaurians, and in reptiles is, at most, a small nubbin of bone lying between the ventral margins of successive centra. In seymouriamorphs the true centra are well formed but are separated by gaps in which moderately reduced intercentra are present.

The embolomeres, even if no longer to be considered as the ancestral amphibians, nevertheless possess, as persistent water-dwellers, many primitive features. Watson in 1926 (199) was able to restore much of the structure of the Carboniferous embolomere *Eogyrinus* (the shoulder girdle, however, is probably incorrectly assigned) and give a description of the skull of *Palaeogyrinus*. Panchen (1964, 1966) has recently redescribed this material. I have collected excellent material of the early Permian genus *Archeria* and have given an account of the girdles and limbs (203); descriptions of the skull and axial skeleton are planned in the near future.

In embolomeres both intercentrum and true centrum form complete central rings. This implies that in the early history of anthracosaurians the intercentrum must have remained persistently large until a time when formation of a ring-shaped "true" centrum had been accomplished by growth of the originally small pleurocentra. Until recently, nothing was known of any antecedent stage in centrum formation. Restudy, however, of *Pholidogaster*, the only described labyrinthodont skeleton from the Lower Carboniferous, shows (Romer 201) that this early amphibian occupied a primitive position in anthracosaurian evolution. As expected from theory, the intercentra are persistently large crescents; the pleurocentra are still paired, but have extended downward to form a pair of thin half-rings. It is significant that in Špinar's larval seymouriamorphs (Špinar 206), the true centra first ossify as two half-rings which later fuse—an interesting ontogenetic recapitulation; I have termed this primitive condition the *schizomeric* stage (Romer 174). In theory, this should be followed by a condition in which the intercentrum is persistently large but the two pleurocentral rings have fused to form a complete, if slender, ring. This stage, which I am terming the *diplomeric*, is at present poorly known in the fossil record; *Diplovertebron* may perhaps belong here. Following this, the embolomerous condition is reached by growth of the intercentrum to form a complete ring, whereas, on the other hand, reduced importance of the intercentrum leads readily to the structural types seen in the seymouriamorphs and primitive reptiles.

It will be noted that the stage which I term diplomeric—in which the pleurocentral ring is complete but the intercentrum is large—is present in *Seymouria* and its relatives, the Seymouriamorpha. But while members of this group have, to a variable degree, preserved the diplomeric vertebral structure, I do not believe that the seymouriamorphs themselves should be placed in a position ancestral to embolomeres and reptiles, since, as early shown by Watson, these forms, in such features as the exaggerated otic notch and the lateral development of the otic region, show group specializations which should not be found in an ancestor of either embolomeres or proper reptiles. Rather, we should consider that once an ancestral diplomere stage was reached, a triple forking took place, with branches to embolomeres (with completion of the intercentral ring), to true reptiles (with intercentral reduction), and to the seymouriamorphs (with retention of the diplomerous vertebra but with development of specializations peculiar to that group).

Ever since Watson's early study of *Seymouria*, the seymouriamorphs have been recognized as of interest as occupying, structurally, a position almost exactly intermediate between advanced amphibians and primitive reptiles. In his classification of 1917 Watson proposed the Seymouriamorpha as one of three suborders of the primitive reptile order Cotylosauria. Once established among the reptiles, *Seymouria* tended to stay there in the minds of various workers, partly because of inertia, partly, I think, because of the desire to have among the most archaic reptiles some type with (in contrast to the *Captorhinus* group) a highly developed otic notch, from which that seen in the presumed reptile *Diadectes* and in typical chelonians could be derived. But certain features of *Seymouria*, such as those mentioned above, are not "desirable" in a reptile ancestor, and the description by Špinar (206) of "branchiosaur" larvae definitely attributable to seymouriamorphs tends to swing the balance to the amphibian side of the divide. The definitive character of a reptile lies, of course, not in its skeletal anatomy but in its amniote mode of development. It is possible that the egg of *Seymouria* began its development in amniote fashion, this followed in the larva by a "lapse" into an aquatic larval phase. But this seems none too probable; and in any event it is becoming increasingly improbable that reptiles, whose divergence from the other anthracosaurians must have occurred in the Mississippian, ever had an ancestral form which would be properly called a seymouriamorphan. Efremov (1946) has advocated the erection of a borderline group, Batrachosauria on the amphibian-reptilian boundary. This is essentially the equivalent of the Seymouriamorpha, as I use this term. But although the seymouriamorphs are close to the boundary, I believe they are a side branch rather than a truly transitional group; the branching out of the reptile stock presumably took place somewhat lower down the anthracosaurian line, in the diplomere stage.

At one time, *Seymouria* appeared to be almost the sole seymouriamorphan. However, a variety of forms has been added with confidence to this group, and further additions may be suggested. A notable contribution was Bystrow's (205) excellent description of *Kotlassia,* one of the last survivors of the group. Additional recently described Russian Permian forms include *Buzulukia*, *Bystrowiana*, and *Chroniosuchus* (Vyushkov 1957). Problematical is the curious flat-headed *Lanthanosuchus* of the Russian Permian (Efremov 1954*b*) which does not fit well into any category of lower tetrapods but is perhaps less a stranger among the seymouriamorphs than elsewhere (although the skull roof pattern is somewhat suggestive of a pareiasaur). Several incompletely known Russian Permian forms which have been assigned to the procolophonoid reptiles seem not improbably to be better assigned here. *Nycteroleter* (Efremov 243) is an animal with a very large otic notch, comparable to that of *Seymouria.* It is described as having the squamosal above the notch, however, and temporal elements are not demarcated. (It seems possible that better material would show the squamosal to have been present in the proper position below the notch, and temporal elements above it.) *Rhipaeosaurus tricuspidens* (not *R. talonophorus*) (Efremov 243), despite dental peculiarities, may be related. *Nyctiboetus* (Chudinov 1955) includes a skull table that appears to be of anthracosaurian type; a bone interpreted as postorbital is apparently an intertemporal.

One further group of supposed reptiles which, on consideration, seem more appropriately placed here is the family Diadectidae, of the early Permian and late Pennsylvanian. *Diadectes* has always been classed among the reptiles (indeed, *Diadectes* is the "type" cotylosaur) and has many reptilian features in its skeleton. But its claim to reptilian status is based on almost exactly the same grounds as is that of *Seymouria*; as both Watson and Olson (1947) have noted, its otic region is *Seymouria*-like; and its enormous and peculiarly developed otic notch cannot have been evolved from any known anthracosaurian or reptilian form except the seymouriamorphs. *Diadectes* should be placed among or close to its seymouriamorphan relatives (Romer 207; Olson 208, 1966). Vaughn (1964) has recently described a *Diadectes* relative possessing the typical otic notch and other characters of the group, but with a normal dentition.

Some years ago Peabody (1958) described as *Hesperoherpeton* a small scapulocoracoid of anthracosaurian type from the Pennsylvanian of Kansas, and Eaton and Stewart (1960) have recently given an account of further materials of this tiny animal, in which the skull fragments, as restored, are difficult to interpret, and the limbs appear to be most unusual in pattern. The vertebrae show the pleurocentra to be in the form of two half-rings, as in Špinar's larval discosauriscids. Eaton believes *Hesperoherpeton* represents a discrete order of amphibians. It may, however, prove, if more complete materials can be found, to be a larval seymouriamorph.

Paleozoic lepospondyls.—Absolutely distinct from the labyrinthodonts, as far as our present knowledge of them goes, was a second series of Paleozoic amphibians, small in size and highly varied in structure, but having as a common character "holospondylous" vertebral centra—basically spool-shaped cylinders with a notochordal perforation. To this assemblage I have applied, in a somewhat expanded form, the Zittel term, Lepospondyli (Baird 214). Nearly a century ago the Miall committee, reviewing such scant knowledge of them as was then available, ranged them in three orders—Aistopoda, Nectridea, and Microsauria—and these terms have continued in almost universal use.

Of the three ancient orders, the Aistopoda and Nectridea have attracted little interest of late although Beerbower (215) has given an excellent account of the "horned" nectridean *Diploceraspis*, Gregory (1948) and the Turnbulls (1955) have described some aistopod materials which add to our knowledge of cranial structure, and Baird (214, 1964*b*) has briefly reviewed the aistopods. The nectrideans, present in the Upper Carboniferous and the base of the Permian, are readily recognizable by the peculiarly constructed caudal vertebrae, in which neural and opposed haemal arches are fan shaped. All known forms have specialized in one of two directions—either toward a long body and tail, with a long slender skull and reduced limbs, or toward a skull structure in which the tabular region on either side of the roof is extended backward as a grotesque "horn." The Aistopoda, confined to the Coal Measures, include only two or three genera with excessively elongated bodies, snakelike in proportions, and with skulls of a very aberrant structure.

More interesting—and more puzzling—is the varied assemblage to which the term Microsauria is applied. The known microsaurs are highly variable in many regards but lack the distinctive structural features which characterize the aistopods and nectrideans. Leaving apart some of the more specialized types, typical microsaurs have general proportions not unlike many modern urodeles and some of the smaller primitive Paleozoic reptiles; since in many cases the structure of small (and often poorly preserved) forms of this sort has been inadequately known, the boundaries and characteristics of the group had been but vaguely defined. In 1950 (209) I surveyed our knowledge of the group and discovered an interesting situation. Although most genera usually included here are holospondylous amphibians, a few prove to be true captorhinomorph reptiles, superficially similar to "true" microsaurs, but readily distinguishable by good diagnostic characters when adequately known. Amusingly, it proves that the genus *Hylonomus*, upon which the group was originally based, is actually a reptile. This gives rise to a problem in nomenclature. One might argue that the group name should follow the "type" genus and be shifted to the captorhinomorph reptiles. But since for nearly a century the concept of the nature of microsaurs has been based on a long series of genera which are amphibian in structure, it has been agreed by workers in the field·(cf. Gregory 212; Gregory, Peabody and Price 210) that common sense dictates a retention of the familiar usage.

Apart from the typical microsaurs, further series of small Paleozoic forms, some at least with considerable body lengthening and with "unorthodox" skull structure, are frequently included among the microsaurs. One is *Lysorophus*, a late Carboniferous and early Permian genus with relatives going back to early Carboniferous times. The body is long (although not as attenuated as in aistopods), the limbs much reduced, and the count of skull elements much reduced. To *Lysororphus* may be allied *Molgophis*, *Metamolgophis* (Romer 1952), and *Palaeomolgophis* (Brough and Brough 1967), with similar body proportions, and the "adelospondyls," comprising the Mississippian genera *Adelogyrinus* and *Dolichopareias* (Brough and Brough 1967; Watson 200) and the recently described *Adelospondylus* (Carroll 1967*b*) from the Pennsylvanian. That these three groups are related to one another is far from certain, and it is still more doubtful whether any of them are really related to the Microsauria proper. The microsaurian skull table differs sharply from that of labyrinthodonts in that there is but a single element to represent tabular and supratemporal, and in the "adelospondyls" even this lone element is restricted to the tabular position, with nothing interposed (despite my onetime hopes to the contrary) between squamosal and parietal.

There is no evidence that any of the Paleozoic lepospondyl groups are related to one another. Further, we are completely in the dark as to their ancestry. If one wished to make out a case for dual (or multiple) origin of amphibians from fishes, the theory should attempt to account for the origin of the lepospondylous groups rather than for the "jump" from crossopterygians to the modern orders of Anura and Urodela, as does Jarvik's theory. Both Westoll (1942*a*) and I (209) have suggested the possible origin of microsaurs from ichthyostegids. But the intervening gap is broad, and the high specialization of these forms in the Carboniferous suggests an antecedent history of considerable length.

The modern orders.—The third discrete chapter in amphibian history is that concerning the three modern orders—Urodela (or Caudata), Apoda (or Caecilia or Gymnophiona), and Anura (or Salientia). Actually the story, as far as we know, is that of two orders only since there are no fossil apodans. There are very few Mesozoic members of either the frogs or the urodeles, and hence the study of these forms is one in which the worker should be in great measure an herpetologist rather than a "pure" paleontologist.

A fair number of additions to our knowledge of fossil salamanders have been made in recent years; much of this work is cited in Estes' recent review (223) and in a compilation on fossil amphibians as a whole by O. Kuhn (1965). For the most part these studies aid in filling out our knowledge of the pedigree of modern families but do not add greatly to the general evolutionary story or tend to modify current concepts of urodele classification to any great degree. The question as to whether in any given case we are dealing with a larval individual or with a permanently larval species complicates the study of fossil urodeles. Westphal (224) has studied giant salamanders and concludes that the living Japanese form *Megalobatrachus* is generically identical with *Andrias*, the famous "Homo diluvii testis"; and fossil cryptobranchids from North America appear also to belong to this genus rather than to the existing American *Cryptobranchus* (Meszoely 1966). Estes noted that *Scapherpeton* and its relatives of the late Cretaceous (Estes 269) and Paleocene are quite similar to the Cryptobranchidae. A sprinkling of Plethodontidae are known from the Cretaceous on, as are Ambystomatidae from the Eocene and later epochs; Salamandridae are rather plentiful throughout the Cenozoic and may have been present in the Cretaceous; in contrast, the Hynobiidae are still absent from the fossil record. Of the long-bodied types, there is an undescribed amphiumid from the Paleocene, and to the Proteidae are doubtfully allotted the oldest urodele remains, *Comonecturoides* from the Jurassic and *Hylaeobatrachus* of the Wealden. *Batrachosauroides* (Auffenberg 1961) is so distinctive a form that it may deserve a separate family.

The Sirenidae have a record dating back to the Cretaceous. Goin and Goin (1962) are so impressed with their differences from other urodeles that they propose separating them as an order Trachystomata. Estes (223), however, believes that the differences are not so great as to demand ordinal separation.

Among the anurans, typical if primitive frogs are now represented by interesting material from the South American Jurassic (Reig 219; Casamiquela 220); considerable good material (e.g., Schaeffer 221) has been obtained in recent years from various Cenozoic horizons, and Nevo has in press a description of new Cretaceous pipid material from Israel and a discussion of pipid evolution. These are being utilized in current studies on frog history by a number of workers, including Hecht (218), Reig, and Tihen (222).[2] Since frog evolution and classification is a complex matter, including not only adult anatomy but developmental history, larval types, and so forth, I feel incompetent to discuss these questions; I have, however, provisionally adopted a division of the Anura into Archaeobatrachia, Aglossa, and Neobatrachia (Reig 219) rather than the

[2] Further recent papers dealing with frog classification and phylogeny are those of Griffiths (1963), Casamiquela (1966), and Inger (1967).

customary divisions in general use since the days of Noble (1931).[3]

We do, however, have one "document" that sheds light on frog ancestry—"*Protobatrachus*" from the Lower Triassic of Madagascar (Piveteau 217; Griffiths 1956). (O. Kuhn [1962] points out that because of a homonym situation, the animal must be known as *Triadobatrachus.*) In this little amphibian, the postcranial skeleton shows no indication of anuran structure, except for some shortening of the column and an elongate ilium. The skull, however, is already of an essentially anuran pattern, and it is reasonable to believe that "*Protobatrachus*" whether a direct progenitor or not, is related to frog ancestry. Hecht (218) is sceptical, and, to be sure, *Triadobatrachus* does lack many of the specialized features of modern frogs. But his position is reminiscent of that of the fundamentalist preacher who said, "They talk about three-toed horses. But everybody knows that a horse has only one toe and if an animal has three toes, it aint a horse." Back of "*Protobatrachus*" the anuran pedigree is as uncertain as is that of urodeles and apodans.

Ancestry of the modern orders.—Until recently all consideration of the pedigree of the modern orders has been based upon the assumption that the anurans, on the one hand, and the urodeles (plus Apoda), on the other, are not at all closely related to one another and, hence, presumably trace back independently to Paleozoic groups.

As regards the urodeles, it has been often thought that the ancestors must lie among the Paleozoic lepospondylous forms. Even here, however, there is an immense time gap of three full geological periods, totaling something on the order of one hundred million years, between the disappearance of the last of the Paleozoic lepospondyls at the dawn of the Permian and the appearance of the first urodeles.

Of the three ancient orders, there is general agreement that the extremely elongate aistopods are not in the urodele ancestral picture. Gregory, Peabody, and Price (210) suggest a relationship between nectrideans and urodeles because of a similar position of the transverse processes of the vertebrae; but the

[3] To my embarrassment, the Aglossa (Pipidae and Rhinophyrinidae) were by mischance omitted from the systematic listings in my text.

known Nectridea are so highly specialized in other features that it is difficult to consider them as ancestors.

The Microsauria, then, remain to be considered. There are no *positive* reasons for deriving urodeles or apodans from members of this group, for typical microsaurs are relatively primitive animals without any of the diagnostic specializations characterizing the modern urodeles, such as loss of dermal roofing elements, palatal modifications, and so forth. A few objections have been raised: for example, the transverse processes of the microsaur vertebrae are, where known, different from those of urodeles (although agreeing with those of the Apoda); the urodele ancestor would be expected to have had an otic notch (Parsons and Williams 216), but none is present in known microsaurs. These are minor objections, however. One can say that microsaurs, in most regards, show a basic structural pattern from which that of the later urodeles and apodans could have been derived.

Parsons and Williams (216) object to microsaurs as salamander ancestors, pointing out that the body, as seen in *Microbrachis*, is too slender and the limbs too reduced for the pedigree of the relatively stoutly built salamanders. But in most cases the postcranial skeleton of microsaurs is inadequately known, and such forms as the gymnarthrids were probably much closer to the heart's desire for urodele ancestors; Cox (1967) suggests that in their general structure the gymnarthrids are not unreasonable as ancestors of the salamanders and other modern orders.

Recently Estes (223), in tabulating seemingly significant osteological characters of salamanders in comparison with various ancient groups, concludes that the statistics favor the temnospondyl labyrinthodonts over the microsaurs by "odds" of 15 to 10. He particularly suggests that the Dissorophidae are related to urodele ancestry. I am not too convinced of the validity of decision by tabulation (nor is Estes himself); the dissorophids are typical members of the "main line" of temnospondyl evolution and show no trend to depart from this in a direction toward salamanders.

I have suggested that the *Lysorophus* group may lie close to the line of ancestry of the Apoda which, like this ancient genus, have assumed a wormlike body shape, and have a skull in which elements present have been reduced in number but secondarily fused together to make a burrowing organ. *Lysorophus* was a water dweller, the apodans mainly burrowers; but one modern genus frequents ponds as well, suggesting a possible change of environment during the history of the group. However, my herpetological friends are decidedly averse to considering this possibility.

Frog ancestry is an equally puzzling problem. Back of "*Protobatrachus*" there lies a major gap. Few have suggested the lepospondyls as ancestors of the Anura, and some authors have favored the labyrinthodonts for this position—particularly because it was believed that the anurans retained traces of labyrinthodont vertebral arches (a point apparently incorrect). For a time a ray of light seemed to appear through Watson's study (1940) of *Amphibamus*. He believed this small Carboniferous form, in which the vertebral centra were poorly ossified, to be a "branchiosaur" and, further, a form in which the cranial roof was undergoing reduction and heading in an anuran direction. But restudy by Gregory (1950) and Carroll (179) shows that, actually (as in the case of other "branchiosaurs"), we are dealing with a normal rhachitomous labyrinthodont, and specifically, a member of the Dissorophidae—a family which shows no positive indication of anuran tendencies.

A recent study by Parsons and Williams (216) suggests a new approach to the problem of the descent of the modern orders. It has always been assumed that frogs and salamanders, so different in many habitus features, are far separated from one another phylogenetically, and hence must have been derived from very different sources. However, it would seem that the easy recognition of obvious adaptive differences between the two groups has blinded us to many significant resemblances. Parsons and Williams point out that the anurans and urodeles are exceedingly similar in a considerable number of characters—in the dentition and auditory apparatus most especially. They possess a number of striking features which are not present in any other tetrapods and which it is highly improbable could have come about by parallelism in evolution. Significant in the modern orders is the development of cutaneous respiration, very recently discussed by Cox (1967). This is a feature which could not have been present in the scaled ancestral amphibians, and is strong evidence for the common descent of frogs and salamanders. It would, thus, appear probable that the two groups are actually rather closely related, and that in our search for the ancestors of anurans and urodeles we are seeking a single ancestral type—not two.

To sum up recent work in amphibian paleontology: we have made great progress in studies of the Labyrinthodontia as regards evolution within the group and as regards their relationships with fish ancestors, on the one hand, and reptiles, on the other. There has been a modest increase in knowledge of the Paleozoic lepospondyls and of the modern orders. But huge lacunae exist, which must be filled before any connected history of lower tetrapods can be told. The older lepospondyl orders resist any attempt to link them with fishes, labyrinthodonts, or the existing orders. And the origin of the modern forms remains as obscure to us today as it was to workers of previous generations.

7

Stem Reptiles

Cotylosaur membership.—Apart from the interest felt in any case in the primitive members of any group, careful consideration of the membership of the Cotylosauria, the "stem reptiles," is unusually important because so much discussion of reptilian evolution and origins has had its roots in their seemingly compound nature. In his classification of 1917, Watson proposed that the cotylosaurs be divided into three suborders—Seymouriamorpha, Captorhinormorpha, and Diadectomorpha. Today the membership in the order, it appears, should be considerably reduced.

As discussed earlier, it seems reasonable that the seymouriamorphs be removed from the Cotylosauria and ranked among the advanced anthracosaurian amphibians. All would agree, however, that the Captorhinomorpha are definitely reptilian. In vertebral structure they have already departed far from the presumed ancestral labyrinthodont condition in that the intercentra are reduced to tiny nubbins. Cranially it is notable that in typical Permian forms there is no trace whatever of an otic notch, although the *Limnoscelis* skull indicates a line of weakness where the notch has been closed. The skull roof pattern is derivable from that of labyrinthodonts, but the intertemporal is never present, the supratemporal is in process of reduction, and tabulars and postparietals are on the way toward elimination from the skull roof, leaving them as elements of the occipital surface. The palate is essentially of a primitive anthracosaurian type, with a moveable basal articulation, and with—a notable reptilian feature—a strong, toothed transverse flange on the pterygoids anterior to the fossa for the temporal muscles. In agreement with anthracosaurians there is a distinct lateral gap, without ossification of any sort, between the sphenethmoid region of the braincase and the otico-occipital segment.

The best-known captorhinids have been, in the past, advanced and specialized Permian members of the group, the Captorhinidae, such as *Captorhinus* and *Labidosaurus*; Olson and others have described further specialized Permian forms. More important, in recent years has been a considerable increase in our knowledge of more primitive and older members of the group. *Limnoscelis* (Romer 240) is obviously a relatively primitive form with, for example, a rather larger supratemporal and a relatively unspecialized dentition. The earliest Permian beds, approaching the Carboniferous boundary, have yielded a series of small captorhinomorphs, such as *Romeria* and *Protorothyris*, as yet incompletely described (Price 1937; Watson 241), which appear to be equally primitive. Still more important, in many ways, is the fact that in "sorting out" some of the members of Carboniferous faunas, a number of Pennsylvanian captorhinomorphs have come to be recognized, such as *Cephalerpeton* (Gregory 212) from Mazon Creek; even in the early Pennsylvanian of the Joggins, a captorhinomorph was already present in *Hylonomus* (Carroll 239). The captorhinomorphs are, thus, a very ancient group, tracing their ancestry back to a point where there are no reptiles and even few amphibians of any sort of greater antiquity.

The third suborder established by Watson was that of the Diadectomorpha, including (1) *Diadectes* and his close relatives, (2) the paraeiasaurs, and (3) the procolophonids. These were characterized by Watson as having the jaw articulation well forward of the occiput and, above it, a large otic notch (as in *Diadectes*) or a "concealed" notch, secondarily covered over by extensions of the cheek bones (as in the other two groups).

In general, the validity of this suborder has not been questioned. But it has become increasingly apparent that *Diadectes*—a relatively ancient and highly specialized type—was, at best, not at all closely related to the other two components of the supposed suborder, which are much later in appearance and lack the curious *Diadectes* specializations. (Note, for example, the isolated position in which *Diadectes* is placed by Watson [242, p. 392].) As discussed above, it now seems clear that *Diadectes* should be removed not merely from the suborder, but completely from the Reptilia, and take its place among the Seymouriamorpha.

Shorn of its "type" member, the "Suborder Diadectomorpha" does, however, appear to remain a well defined and essentially coherent group of primitive reptiles, which may perhaps be termed the suborder Procolophonia. In their end forms, procolophonids and pareiasaurs have assumed very different adaptive features, but the little Russian *Nyctiphruretus*, although assigned to the procolophonid group, appears to be not far removed structurally from a position of ancestry to pareiasaurs and procolophonids. In many regards, *Nyctiphruretus*, except for the loss of the supratemporal, shows a skull pattern not too different from that of the captorhinomorphs.

There is, however, one conspicuous difference—namely, that the jaw is shortened, the articulation hence is well forward, and the cheek is correspondingly modified, its back border slanting forward and downward from skull roof to quadrate. This back border, in *Nyctiphruretus*, is somewhat concave. This margin may well have been utilized to aid in supporting a tympanic membrane, and can thus, in a sense, be called an otic notch. But it seems clear that it is primarily brought into existence as a result of jaw shortening, that it has no genetic connection with the old-fashioned otic notch of the labyrinthodonts.

Of the original components of the Cotylosauria there thus remain the Captorhinomorpha and the Procolophonia (the reduced Diadectomorpha). The two differ mainly in that the procolophonians are rather more advanced and specialized in certain regards, notably the shortening of the jaws and consequent reshaping of the cheek.

To these remains of the original constituents of the order, certain other forms may well be added. Although reptilian classification has historically been based on temporal openings, and the cotylosaurs primarily defined by the absence of such openings, it must be kept in mind that, after all, this is merely a useful means of ready identification of types, and that there may well have been a variety of short-lived lines arising from the cotylosaurs in which temporal openings were developed independently of similar happenings in the ancestry of major groups. A form in this category is *Bolosaurus* (Watson 241). Here there is a small temporal opening far down on the side of the cheek. Apart from this, and the presence of peculiar bulbous teeth, *Bolosaurus* is a good cotylosaur and—despite the peculiar dentition—appears to be essentially a somewhat aberrant captorhinomorph, with no "diadectomorph" affinities.

Vaughn (355) would place *Araeoscelis* in essentially the same relationship to the captorhinomorphs. It is here discussed in relation to the euryapsids.

Possibly best placed here is a group of small forms from the South African Permian—the "millerettids" such as *Milleretta*, *Millerosaurus*, etc.—first described by Broom (394) and later discussed by Watson (242); with them Watson would associate *Mesenosaurus* of the Russian Permian. Watson considers these animals as ancestral to the diapsids, and with this in mind, I brigaded them in my *Osteology* with the eosuchians. Watson currently proposes that they be placed in a separate order Millerosauria; a position as a subgroup of the Cotylosauria is perhaps appropriate. For the most part, knowledge of the millerettids is confined to the skull. In much of their structure, differences from captorhinomorphs are not marked, and the resemblance is even closer than in the Procolophonia, in that the supratemporal is retained. The quadrate position is variable—the jaw articulation is, for example, well to the back in *Mesenosaurus*, somewhat farther forward in *Milleretta* and *Millerettops*, well forward in *Millerosaurus*. In general there is some degree of concavity at the back border of the cheek, suggesting, as in *Nyctiphruretus*, initiation of a tympanic support in an "otic notch." The one unusual feature is the trend for development of a lateral temporal opening in a somewhat pelycosaurian fashion. The cheek is completely covered in *Millerettops*; in *Milleretta*, jugal, postorbital, and squamosal barely fail to meet; in *Millerosaurus* and *Mesenosaurus* there is a well-developed and well-defined lateral temporal opening.

I have already commented, in the case of the term Microsauria, on the chaos which might be caused were the systematists ever to extend their rigid rules to the names of higher categories. "Cotylosauria" is another case in point. For two-thirds of a century this term has been synonymous with "stem reptiles." But, if one traces back, the "type" cotylosaur was *Diadectes* [*Empedocles*] (Cope 1880), which now appears to be a seymouriamorphan. If then, one insists that the name follows the type, the term Cotylosauria would (absurdly!) be transferred to cover the seymouriamorphans!

Apart from a number of papers already mentioned, various additions to our knowledge of cotylosaurs have been made in recent years. New finds, although fragmentary, have been made of *Limnoscelis* relatives in the early Permian of New Mexico and Colorado, and the assumption that *Limnoscelis* of the Permian is an archaic "holdover" of a primitive group is reinforced by the description by Carroll, now in press, of a Carboniferous form from Nova Scotia. A number of more advanced captorhinomorphs have been described from the American early Permian, such as *Paracaptorhinus*, *Captorhinikos*, *Captorhinoides*, *Kahneria*, *Labidosaurikos*, *Melanothyris*, and *Rothioniscus*; and the group survived in the Russian Permian in *Geocatogomphius* and *Hecatogomphius* (Olson 383). Little has been done recently on pareiasaurs, except for the description and discussion of forms from the Russian Permian—*Leptorhopha*, *Parabradysaurus*, *Rhipaeosaurus*—which appear to be primitive members of this group (Chudinov 1957). Considerable of value has been published on procolophonid types; Colbert (245), in describing new material of *Hypsognathus* gave a general review of our knowledge of the group, and a number of new forms have been described, including *Candelaria* from Brazil, *Neoprocolophon* and *Paoteodon* from China, *Phaanthosaurus* and *Tichvinskia* from Russia, *Barasaurus* from Madagascar, and *Sphodrosaurus* from North America.

Cotylosaur ancestry.—The ancestry of the "stem reptiles" has been much discussed. Because of the seeming close relationship, structurally, between reptiles, on the one hand, and seymouriamorphs and the labyrinthodonts of anthracosaurian type, on the other, the general opinion has been that reptiles sprang from labyrinthodont ancestors. This is, I think, still the common opinion at the present time, although the transfer to the amphibians of *Seymouria* and *Diadectes*, the only supposed reptiles with a highly developed otic notch of primitive type, removes one point of supposed resemblance. As pointed out above, the curvature of the back border of the cheek in primitive procolophonians (and millerettids) is hardly comparable to the original otic notch, since it is situated well down the side of the cheek, in strong contrast to the dorsally incised labyrinthodont notch. We are essentially forced to the conclusion that the ancestry of reptiles lay among the captorhinomorphs, in which the notch is absent, and in most known forms roof and cheek are tightly sealed together in its former position. It is quite possible, as I discussed at one time (240), to envisage the closure of the original otic notch, with migration of the ear drum downward toward the quadrate region, and attainment of the captorhinomorph condition from that of an ancestral anthracosaurian. But, as we have seen, the captorhinomorphs are a very ancient group, and there are no records of such an evolutionary process which (if it took place) cannot have occurred later than about the Mississippian-Pennsylvanian boundary.

It is thus easy to see why (even when the difficulties which are apparent today were not fully realized) there have from time to time been suggestions that part or all of the reptiles were derived from some other source—that source being the microsaurs of the Carboniferous.

Leaving aside the obscure Lower Carboniferous forms and the highly specialized lysorophids, we see in the "typical" microsaurs small animals with a rather normal body build and a skull which superficially resembles that of a captorhinomorph. In 1950, as mentioned earlier, I attempted to sort out the group and, as far as could be done with generally poor material, to discover distinctive features by which microsaurs and captorhinomorphs could be distinguished (Romer 209). Skull differences are definite, and the typical microsaurs are long-bodied, feeble-limbed forms, with a reduced pectoral toe count—characters hardly expected in an ancestral reptile. The Baird-Carroll paper mentioned earlier gives additional diagnostic features which will be sufficient, I fondly hope, to lay the microsaur ghost as regards potential reptile ancestry.[1]

Westoll in 1942 (1942*a*) suggested that the microsaurs were ancestral to all true reptiles. Olson in 1947 advocated descent of diadectomorphs and chelonians from seymouriamorphs and other reptiles from microsaurs, but he has since modified his views. Tatarinov in 1959 proposed that "sauropsids" were of

[1] Brough and Brough (1967) have interpreted a few (a very few) small structures on the electrotype of the microsaur *Microbrachis* as dorsal intercentra, whereas I have argued that a distinctive character of microsaurs, as contrasted with primitive reptiles, is the absence of such structures. However, it is far from certain that these structures are actually intercentra. They are not improbably broken pieces of the ventral rim of the adjacent centrum. (This is the case, for example, of a supposed intercentrum in the microsaur *Ostodolepis* figured by Williston [1914, Fig. 31].)

labyrinthodont descent, "theropsids" (essentially the synapsids) derived via captorhinomorphs from microsaurs. Vaughn (1960) advocated a similar polyphyletic origin for reptiles, and in 1962 argued that captorhinomorphs and microsaurs are, at least, related, and most recently Brough and Brough (1967) have argued for microsaur-captorhinomorph relationship. Kuhn-Schnyder (1963), although not going into details, suggests that the reptiles are not a true group but simply an evolutionary "grade," and that perhaps as many as six different reptile groups have evolved upward independently from the amphibian level.

First, as regards polyphyletic origin of reptiles, I think that no one who broadly considered all the factors involved would seriously think it a possibility. After all, a reptile is defined as a tetrapod that has attained the amniote type of development. This includes a very complicated series of embryological processes, with the formation of amnion, chorion, allantois and yolk sac—processes basically similar in all living reptile groups and in birds and mammals as well. It is practically inconceivable, as mentioned earlier, that the descendants of two or more separate groups of amphibians could have independently achieved the development of this whole series of structures. Polyphyletic origin need not be considered.

As to the microsaurs possibly being ancestral to reptiles—and I am sceptical of this—it may be pointed out that acceptance of this theory would get us nowhere as to the ultimate derivation of the reptiles. Advocacy of microsaur ancestry is underlain, apparently, by the assumption that the microsaurs are an older group than the captorhinomorphs. They are not. Apart from the Scottish Lower Carboniferous genera, whose relationships to the Microsauria proper are, as already mentioned, very doubtful, the oldest known microsaurs are those of the Joggins tree stumps, and in the very same fauna are the oldest captorhinomorph remains. If captorhinomorphs and microsaurs are related, it is as reasonable (or unreasonable) to argue that microsaurs are descended from captorhinomorph reptiles as to believe the reverse.

Reptile classification.—Discussion of reptilian phylogeny and classification will in general be deferred here for consideration under the various groups concerned, but a few notes may be made at this point. The history of the subject was reviewed in my 1956 *Osteology of the Reptiles* (225). For the most part, the growth of schemes of classification and evolutionary differentiation has been based on a terminology derived primarily from temporal arch structure. This type of classification was carried to a high point by Williston. In my own work I have tended to maintain a conservative position, using in part a temporal arch terminology, but with recognition that arch structure is only one element in the situation. In some major groups (such as placental mammals), there appears to have been a wide, early diversification, with development of a large number of orders not readily "clumped" into superordinal assemblages. Among reptiles, on the contrary, a number of major groupings are apparent. The customary association of cotylosaurs and chelonians as a subclass Anapsida is somewhat artificial but harmless. It is universally recognized that the Synapsida, a great group of forms leading eventually toward mammals, is a natural assemblage. Despite Williston's efforts to include them in a group of Parapsida (an assemblage surely artificial), the ichthyosaurs have (up to now) remained stubbornly independent. (But see the discussion in chapter 8.) A large proportion of reptiles, living and fossil, are diapsids with two temporal fenestrae. It is, however, highly improbable that all diapsids are descended from a common ancestor, and the array of diapsids is so vast and varied that, apart from this, it has seemed to me better as a practical matter to split them into two subclasses: the Lepidosauria, including more primitive forms and Squamata, and the Archosauria, the great group of the dinosaurs and their associates. The one problematical assemblage is that of forms with an upper temporal opening, with the sauropterygians the main component—a group Williston termed the Synaptosauria but for which Euryapsida (Colbert 4) seems preferable. We thus have (apart from euryapsid and ichthyosaur perplexities) only a few major groups, all of which may reasonably be considered to have essentially discrete lines of descent from the ancestral forms—which, as far as I can see, may well be the ancient captorhinomorphs.

Watson (241, 242), however, revived a suggestion made long ago by Goodrich that all reptiles be classed in two groups,

Theropsida and Sauropsida, the former being essentially the mammal-like reptiles, the latter including nearly all other members of the class. Goodrich based his argument mainly on the basis of aortic arch construction, in which mammals show a pattern basically different from that found in existing reptiles. Watson uses the auditory apparatus as his major basis of distinction. He believed that the "theropsid" line consists of captorhinomorphs and synapsids (and perhaps euryapsids and the ichthyosaurs); the "theropsids" are assumed to be derived from seymouriamorphans and to lead to later major reptile groups by way of procolophonians and millerettids (cf. the phylogenetic table in Watson 241, pp. 392–93). Watson assumes that the massive stapes found in most "theropsids" could not function in transmitting vibrations from a tympanum, and that under these conditions and with no formed otic notch, an eardrum did not reappear in "theropsids" until an advanced stage was reached. "Sauropsids," on the other hand, are believed to have had an otic notch and tympanum throughout their history.

I cannot convince myself that this thesis can be maintained. Parrington (1958) has pointed out various strong objections. As he and Hotton (1959, 1960), among others, have suggested, an eardrum may well have been present in "theropsids" despite the absence of a well-formed otic notch, and even a stapes as massive as that in pelycosaurs is competent to transmit vibrations. And on the "sauropsid" side of the story, little evidence appears to exist of the continuous presence of a well-formed otic notch from ancestral types on to "modernized" forms (a point which I am discussing in a paper now in press; Romer 1967*b*). The definite reptilian otic notch is essentially a structure placed low down on the side of the cheek behind the quadrate. It is quite different in position from the primitive notch, which was more dorsal in position. And there is no known connecting link between the seymourian notch and that of "modernized" forms; in such supposed connecting links as primitive procolophonians and the millerettids, there is at the most a gentle concavity of the cheek contours. It seems more reasonable to believe that we are here witnessing the first stages in the development of a new type of support for the tympanic margin.

Huene (226), we have noted earlier, in 1956 evolved a new

classification of lower vertebrates which at first sight appears to differ radically from that commonly accepted; I am discussing it elsewhere (Romer 1967*b*). As regards reptiles, he originally believed, on the *Tupilakosaurus* evidence, the ichthyosaurs to have evolved independently from amphibians, a view he later abandoned. He combined turtles and cotylosaurs with certain relatives, or supposed relatives of the reptiles, in a "Ramus Reptiliomorpha," which is essentially the order Cotylosauria. All remaining reptiles are divided into two groups, Sauromorpha and Theromorphoidea. These correspond in general to Watson's Sauropsida and Theropsida, discussed above.

Persistence of primitive aquatic habits in reptiles?—Because of their possession of an amniote type of development, the reptiles, in contrast to their necessarily amphibious-to-aquatic ancestors, were well qualified to take up a purely terrestrial existence, and many of them rapidly became typical land dwellers. When, therefore, we find reptile types which are aquatic in habits, we tend to assume that water-dwelling is secondary and that such forms were descended from purely terrestrial ancestors.

This is not necessarily the case. As I have pointed out elsewhere (1957, 1958), the development of primitive reptiles with an amniote type of reproduction occurred before there was any proper supply of food on land for types which appear to have fed originally only on animal matter. Some early cotylosaurs, such as *Limnoscelis*, appear to have been primarily amphibious in their mode of life, and this is also true of the most ancient and primitive pelycosaurs of the ophiacodont group.

Not improbably, the ancestors of certain other amphibious-to-aquatic reptiles never truly gained the land to revert to water. For example, few chelonians are purely terrestrial, and most of the older members of the order appear to have been aquatic to a greater or lesser degree—perhaps primitively so. Among the archosaurs, many of the dinosaurs were presumably terrestrial—as were, of course, the bird and pterosaur ancestors. But part, at least, of the herbivorous dinosaurs—sauropods and duckbills—were definitely amphibious; so are the crocodilians from the beginning of their history, and so were the phytosaurs, prominent in the ancestral archosaur group of the Thecodontia. Probably the ancestral archosaurs were primarily amphibious.

About the most prominent water reptile groups, the sauropterygians and ichthyosaurs, there is insufficient evidence regarding ancestry; but it is far from impossible that the ancestral forms were, to at least some degree, primitively amphibious.

8

Varied Reptilian Types

Chelonia.—Relatively little noteworthy material has been published on fossil turtles during the two decades elapsed since the publication of the second edition of the *Vertebrate Paleontology.* Among major items, I note that Zangerl (251, 254) published important accounts of Cretaceous chelonians; Bohlin (1953) described a number of fossils from Kansu and Mongolia; various papers by E. E. Williams during the 1950's are important in chelonian classifications and relationships; Parsons and Williams (1961) have given a valuable morphological description of Jurassic skulls; Bergounioux has described as new, in several papers, a number of European and African Tertiary specimens.

None of the recent publications has tended to change current concepts of the general evolutionary picture of the group. Half a century ago there was much interest in the curious leathery turtle *Dermochelys* [*Sphargis*], in which the carapace consists only of isolated denticles, and theories were developed in which this condition was thought to be primitive, the shell developing at a relatively late stage of chelonian evolution. Most later writers have abandoned this viewpoint in favor of one in which the developed shell, as seen in the oldest well-known type, *Proganochelys* [*Triassochelys*], is primitive. A majority of relatively primitive Mesozoic chelonians are regarded as forming a primitive suborder Amphichelydia, from which sprang the Cryptodira of the Cretaceous and later times as a progressive continuation of the main evolutionary line, with Pleurodira as an aberrant but structurally conservative side branch; *Dermochelys* is regarded as a specialized derivative of the marine cryptodires, and the leathery turtles, *Trionyx*, a further cryptodire side branch in which the horny scutes have been reduced. I had adopted a classification of this type in my *Osteology of the*

Reptiles and in my last edition of the *Vertebrate Paleontology*, and am continuing it in the third edition.

Huene's system of classification (226, pp. 199–220) is none too clear; he does not use the term Amphichelydia but instead uses Cryptodira for the main evolutionary line from Triassic to Recent, placing in the pleurodires a number of families which are here included in the Amphichelydia, and making a separate suborder for the *Trionyx* group. Bergounioux (1955) uses a classification reminiscent of the thinking of a half-century ago, in that he divides the turtles into two suborders: one, the "Athèques," including only the leathery turtles; the other, the "Thécophores," all remaining turtles. The latter are divided dichotomously into one superfamily including only the *Trionyx* group, and a second superfamily of "Lépidodermes" in which are lumped all other turtles, from the most primitive forms on. This classification gives very little suggestion of the probable evolution of the group. Bergounioux (1938) many years ago described what clearly appears to be a familiar type of septarian nodule from the Permian under the name (fortunately preoccupied) of *Archaeochelys*; he continues to list this in his 1955 account (p. 505).

Watson many years ago pointed out features of *Eunotosaurus* of the Middle Permian of the Karroo that suggest chelonian relationship. The animal is, however, so far removed in so many characters from the true chelonians that herpetologists shudder at the thought of it (cf. Parsons and Williams 1961). It is currently being restudied by Cox.

It has always been assumed that, as proper anapsids, the turtles have sprung directly from some cotylosaurian stock. The Diadectomorpha of Watson's classification have seemed the logical point of origin, mainly because of the prominence of the otic notch in chelonians and its development, to a variable degree, in this group. Olson (1947) suggested relationships to the Diadectidae in particular and advocated a dichotomy of all reptiles into two unequal groups, the Diadectomorpha and Chelonia comprising the Parareptilia in contrast to all other members of the class. In Watson's proposed cleavage of the reptiles into Sauropsida and Theropsida, the chelonians, with diadectomorphs, are integral parts of his sauropsid series. Gregory (1946), like Olson, compared chelonians and diadectomorphs, but his comparison was made with the pareiasaurs rather than with the diadectids.

If, as I believe, *Diadectes* is a sterile end-form of the Anthracosauria, this animal may be counted out as a turtle ancestor; *Diadectes*, after all, is highly specialized and has little in common with chelonians except the presence of a large otic notch. But there is no reason to exclude the remaining members of the erstwhile "diadectomorphs" from potential chelonian ancestry.

The figure of the skull of "*Triassochelys*" (=*Proganochelys*) as given by Jaekel (250) is quite misleading. The new material, to be described by Staesche, shows a number of conspicuous differences; these can be seen in the photographs given by Parsons and Williams (1961) and in the restoration in my text (*Vertebrate Paleontology*, Fig. 166) based upon their figures and information given me by Williams. In palatal structure, particularly, it is closely comparable to the procolophonian *Nyctiphruretus*, and a derivation of the chelonians from the "Diadectomorpha"—as now reduced to the Procolophonia—seems reasonable, thus confirming to a considerable degree the beliefs of Gregory and Olson about their pedigree.

Mesosauria.—No new light has been shed in postwar years on *Mesosaurus*. Huene's material (371) suggests the presence of a lateral temporal opening in the skull, somewhat comparable to that of synapsids. This is none too certain, but even if this interpretation is correct, it by no means proves a close relationship to synapsids, and nothing else in the structure of this curious little creature is particularly comparable to synapsid structure. Obviously the specializations seen here in the early Permian in skull, dentition, and limbs indicate a considerable period of evolution of this amphibious type from, presumably, ancestral captorhinomorphs, well back into the Carboniferous. When, if ever, a pre-Permian vertebrate record is discovered in the southern continents, the story will presumably come to light. Ginsburg (1967) has suggested that *Mesosaurus* is related to the sauropterygians, but this suggestion has little basis other than the fact that *Mesosaurus* is antecedent in time to the nothosaurs and is somewhat similar in showing aquatic adaptations. Despite the presumed presence of a temporal opening, it seems best to consider *Mesosaurus* as a short twig from the cotylosaur stock and to retain it within the subclass Anapsida.

A major item of interest with regard to *Mesosaurus* is paleographical; its known presence only in western South Africa and southern Brazil is an argument in favor of apposition of South America and Africa in late Paleozoic times. It has been suggested that its absence in other early Permian or late Carboniferous deposits may be associated with its unusual feeding habits and presumably specialized ecology, and that it may after all have been widely distributed. Perhaps; but lack of evidence of this creature in other late Paleozoic deposits, despite the very considerable amount of work done on them in recent decades, makes this suggestion increasingly improbable.

Ichthyosaurs.—Up to the time of publication of my third edition little advance had been made in recent years in our knowledge of ichthyosaurs. More knowledge is badly needed about the oldest-known ichthyosaurs of the Triassic—the skull, particularly, is very imperfectly known; Camp (1954) discovered an interesting mass "graveyard" of giant Triassic individuals, but no scientific data have as yet come from this. We still lack any knowledge of ichthyosaur ancestors in the earlier Triassic or the Permian. I pointed out some time ago (Romer 1948) that, the temporal opening apart, such primitive pelycosaurs as the ophiacodonts show structures which could well be antecedent to those of the ichthyosaurs, and it is reasonable to believe that they sprang from "cousins" of the primitive pelycosaurs among the generalized captorhinomorphs, from which the ophiacodonts have not departed far in structure.

Interest was aroused by Nielsen's description in 1954 of *Tupilakosaurus* from the early Triassic of Greenland. The type included a partial skull table and vertebrae with thin but complete ring centra. Although Nielsen thought (erroneously, it would seem) that there were two such centra to each segment (an embolomere structure), the shape of the centra suggested that of the ichthyosaurs and hence advocated the possible descent of the ichthyosaurs from such a form. This suggestion was seized upon by Huene (226). The skull pattern, as far as preserved, resembled that of early Triassic temnospondylous amphibians, and Huene would have the ichthyosaurs descended from labyrinthodonts quite independently of all other reptiles. Huene, however, later retreated from this position (1959, 1960).

The true situation has been revealed by the description by Shishkin (1961) of a specimen from Siberia with a complete skull attributed to *Tupilakosaurus*. The skull is that of a typical advanced temnospondyl of brachyopoid type, showing not the slightest indication of ichthyosaur characteristics. The centra are of the sort developed from the intercentrum in a few advanced stereospondyls.

Appleby (375) has described the cranial morphology of *Ophthalmosaurus* and in 1961 suggested a relationship between ichthyosaurs and chelonians. The points of resemblance noted, however, appear to be much more than counterbalanced by the extremely divergent adaptations of the two groups.

In the usual description of the temporal region of ichthyosaurs (as, for example, in the figures of *Ophthalmosaurus* by Andrews [359] and Gilmore [373]), there is said to be present a lateral element identified as a squamosal lying above a small quadratojugal and behind the postorbital. In consequence, the large bone present farther dorsally, forming much of the boundary of the temporal opening and supporting the quadrate, has necessarily been termed a supratemporal despite the fact that it has the proportions and relations of a squamosal. Thus bounded, the ichthyosaur temporal opening appeared to be very different from that of any other reptiles; this situation was responsible for the ichthyosaurs being placed in a discrete reptilian subclass.

In a Cretaceous ichthyosaur skull which I am currently studying, however, I found, to my astonishment, no trace of the supposed lateral "squamosal"; instead, a large quadratojugal extended far up the cheek. Either this Cretaceous form is aberrant, or the customary description of the ichthyosaur cheek is perhaps incorrect.

The latter appears to be the case. In Triassic ichthyosaur skulls the cheeks are often poorly preserved, but invariably a lateral "squamosal" is assumed to be present. But in the skull sectioned by Sollas (374) there was, he notes, no such element, although all other structures in this region were well preserved. This past summer I had the privilege of studying in England three Liassic skulls prepared by acid; in all three there was—

as in my Cretaceous skull, and in contrast to the usual description—a large quadratojugal and no trace of the supposed "squamosal." The Andrews and Gilmore figures are quite accurate, but it appears that the supposed "squamosal" is merely the upper part of the quadratojugal, the middle part of which is covered by the postorbital![1]

If, as appears to be the case, the supposed lateral "squamosal" is nonexistent, the large dorsal element now termed supratemporal becomes the true squamosal. And in consequence the temporal opening is essentially identical with the euryapsid type present in that very different aquatic group, the Sauropterygia.

[1] I am informed that Mr. Christopher McGowan, studying part of this material under Prof. Attridge, has independently made the same discovery.

9

Euryapsid Reptiles

Euryapsid early history.—Williston clearly perceived that the plesiosaurs, nothosaurs, and placodonts were a discrete subgroup of reptiles, with an upper temporal opening of "normal" type an obvious key character. Believing that the presence of a single temporal arch and opening implied some relationship to the Synapsida, he coined for them the rather awkward subclass term Synaptosauria. Colbert (4), after discussion with me and others, reasonably suggested Euryapsida as a substitute.

Placodonts and nothosaurs were already characteristically developed by the Middle Triassic, and the plesiosaurs obviously stemmed during this period from nothosaurlike forms. But whence did this group arise? Presumably from captorhinomorph ancestors, ultimately; but what evidence have we of their pedigree in the Permian or early Triassic? (I noted earlier that a suggestion of *Mesosaurus* relationships rests on no adequate basis.)

The only adequately known Permian form with an euryapsid temporal opening is *Araeoscelis* (with which *Kadaliosaurus* is probably identical generically), as originally described by Williston and later thoroughly discussed by Vaughn (355). With *Araeoscelis* Williston associated *Protorosaurus* and several other Permian and Triassic forms to constitute an order Protorosauria. I have tended to gain the concept of there having been during Permian and Triassic days a considerable radiation of varied euryapsid types, mainly of small size, which were, however, replaced in the main by lepidosaurs and archosaurs, although leaving the placodonts and sauropterygians as successful aquatic survivors.

I must confess, however, that this concept is none too secure. Surely there were at least some Permian and early Triassic euryapsids to afford ancestry to the successful aquatic groups.

That there was at least considerable variety among primitive euryapsids is shown by *Araeoscelis*, a slenderly built early Permian form of rather primitive build, and the very different *Trilophosaurus* of the Triassic (Gregory 357). I have fancied a number of small and obscure Permian and Triassic forms (such as *Coelurosauravus*, *Weigeltisaurus*, and *Trachelosaurus*) to be members of the early euryapsid radiation; but in default of adequate knowledge (particularly of the temporal region) it is difficult to be sure whether we are dealing with euryapsids or early lepidosaurs. I have thought that some of the odd types whose remains have been found by Peyer and his associates in the Middle Triassic, such as *Macrocnemus* and *Tanystropheus*, might be members of that group. It now proves, however, that *Macrocnemus* is a primitive lepidosaur (Kuhn-Schnyder 1962, etc.). The skull of the grotesquely long-necked *Tanystropheus* is still imperfectly known; whether it is a lepidosaur or a euryapsid is still uncertain.

Williston placed early forms which had, or which he thought to have, a temporal opening of euryapsid type in an order Protorosauria, using *Protorosaurus* of the late Permian of Europe as the "type" genus. The skull of this form is (again) poorly known, and we cannot say whether it is a euryapsid, as I fondly believe (Romer 1947), or a lizard relative, as Camp (259) maintains. An x-ray study of the Royal College of Surgeons specimen is strongly indicated. Under the circumstances, it seems inadvisable to use this doubtful form's generic name in ordinal fashion, and I therefore, following the lead of certain earlier workers, use Araeoscelidia as the ordinal name for primitive euryapsids.

In my concept of the probable radiation, in the later Permian and Triassic, of a series of primitive euryapsids with a single upper temporal opening, I formerly assumed that the sauropterygians and placodonts were the only survivors of this radiation, otherwise "swamped" by the rise of lepidosaurs and archosaurs. But my very recent discovery, mentioned above, that the ichthyosaurs have a typically euryapsid temporal opening raises new questions. The ichthyosaurs are, like the sauropterygians, aquatic types; but the marked differences in structure and presumed mode of life make it clear that there is no intimate connection between the two. Either ichthyosaurs and plesiosaurs represent two lines which diverged at the very dawn of euryapsid history or, alternatively, the euryapsid temporal opening evolved more than once.

Kuhn-Schnyder (1961, 1963) maintains that the sauropterygians are not at all related to the placodonts and have nothing to do with the euryapsids, but are of diapsid derivation, with an independent evolutionary line extending back to an amphibian stage. He points out certain seemingly primitive features in the nothosaur *Simosaurus*, such as the large tabular (much as in pelycosaurs), and bases his belief of a previous diapsid condition on the slender, arched zygomatic bar with a reduced quadratojugal, and on the jaw articulation being placed well below the arch. The ventral position of the articulation, however—functionally advantageous in certain types of feeding habits—is not peculiar to sauropterygians; it is seen, for example, in many synapsids, in placodonts, and in ornithischians. Further, the condition of the zygomatic arch seen in sauropterygians by no means proves the former presence of a lateral temporal opening and lower arch. The Chelonia are anapsids, in which surely no lateral temporal opening or lower bar was ever present; but a series of turtle skulls (e.g., in my *Osteology of the Reptiles*, Fig. 48) shows that the temporal region can be excavated from below to as great—or even much greater—an extent than in any sauropterygian. Kuhn-Schnyder believes the sauropterygians evolved independently from labyrinthodonts, but there seems to be little positive evidence for this point of view.

Sauropterygians.—The main stock definitely assignable to the Euryapsida is that of the Sauropterygia, including the plesiosaurs of the Jurassic and Cretaceous and their Triassic relatives and predecessors, the nothosaurs. No striking new discoveries have been made recently among the plesiosaurs but I may cite as substantial contributions descriptions of various American forms by Welles (360, etc.), a review of the pliosaurs by Tarlo (1960), a reconstruction of the skeleton of the giant *Kronosaurus* (Romer 1959), a review of Cretaceous plesiosaurs by Welles (1962), and a general review of plesiosaurs by Persson (364). Due to the usually crushed condition of the materials, our knowledge of the structure of plesiosaur skulls is still far from adequate.

During the prewar period, very considerable additions were made to our knowledge of nothosaurs by discoveries in the Monte San Giorgio excavations, initiated by Peyer and now being continued by Kuhn-Schnyder. The latter has described further nothosaur remains (1960, 1961, etc.), and Huene in 1952 (1952*a*) gave a description of the skeleton of *Simosaurus*.

Placodonts.—In my *Osteology* I included the Placodontia in the order Sauropterygia. On reconsideration, I feel that this is improper, and that these interesting Triassic mollusc-feeders should be placed in a separate order. I believe, however, it is still true that placodonts and sauropterygians are rather closely related (despite the claim by Kuhn-Schnyder, discussed above, of wide separation). In the decade preceding the war, our knowledge of the group had advanced considerably, with, for example, the description of a complete skeleton of *Placodus* and Huene's discovery of the highly specialized "pseudo-turtle" *Henodus*. The most important find since is the description by Peyer (366) of *Helveticosaurus*, a primitive form in which the typical dentition of the placodonts had not yet developed. Kuhn-Schnyder (365) has recently given a general survey of the group as well as a study of the *Cyamodus* skull (1959).

10

Lepidosaurian Reptiles

The lizards were long thought by Williston and others to be forms which had never possessed more than a single temporal arch and opening, and were ranged by Williston in awkward relationship to the ichthyosaurs in a subclass Parapsida. It was, however, pointed out by Broom (257) forty years ago that they were actually diapsid derivatives in which the lower temporal arch had been lost. Consequently, it became apparent that they were closely related to the rhynchocephalians (rather than as previously supposed, members of a distinct subclass) and, together with the rhynchocephalians and a few obscure Mesozoic forms, appeared to constitute a rather coherent group of diapsids. But they are a group which, apart from the primitive common possession of two temporal openings, show absolutely no similarity to the great diapsid group of archosaurian reptiles and, consequently, should not be lumped with them in an artificial "subclass Diapsida." I hence revived for them, in the first edition of my *Paleontology*, the term Lepidosauria, once used for lizards and *Sphenodon*, and utilized this in the sense of a subclass. This usage has, I think, been very generally accepted.

Eosuchians.—But although the lizards and snakes, the Squamata, and *Sphenodon* and its fossil allies, the Rhynchocephalia, make up the greater bulk of the membership of the subclass Lepidosauria, we find in the Mesozoic, and particularly in the Triassic, a modest number of two-arched forms which do not belong to either of the two surviving orders and show in some cases a primitive structure of a sort one would expect in the common ancestors of the two major groups. For these, Broom's ordinal term Eosuchia can be used (despite the misleading nature of the name).

Primitive here is *Youngina* of the Upper Permian of South Africa, with which can be associated a few closely related (if not generically identical) forms such as *Younginoides* and *Youngopsis* as well as, provisionally, a considerable number of poorly known little reptiles from the late Permian and early part of the Triassic. Members of the *Youngina* group have fully preserved the lower temporal arch, as in rhynchocephalians, but on the other hand do not have the specialized acrodont dentition of *Sphenodon* and its allies. Their structure and time of appearance suggest that we have here truly primitive and ancestral lepidosaurs from which both lizards and rhynchocephalians may have taken origin. As pointed out earlier, they may well have arisen from the millerettids, which might be included in the Eosuchia or considered, as I have in my present text, as advanced cotylosaurs probably ancestral to the eosuchians.

In the Triassic as well we find a few forms such as *Pricea* and, notably, *Prolacerta* (Parrington 258; Camp 259), which are clearly derivable from *Youngina* or its allies but are in the process of losing the lower temporal arch (and the quadratojugal element with it) and strongly point toward the Lacertilia. How should these forms be classified? As presumable ancestors of the lizards, they might be considered a basal group of the Squamata, as Tatarinov has done in the Russian Treatise (1964) or, preferably, I think, be treated as advanced eosuchians, as I have done in my text and as Piveteau (1, vol. 5, pp. 545–55) has done in the French Traité. In the Triassic, increased knowledge of small reptiles of the time is tending to make us acquainted with a series of forms showing all transitions from typical eosuchians to typical lizards. Where shall we draw the line between the two orders? Dr. Pamela Robinson, in conversations with me and in a paper now in course of publication, has suggested that we include in the Squamata only those forms in which the typical lizard temporal region, with a fully moveable quadrate, has been evolved. Under such terms, *Prolacerta* and its allies should be retained within the Eosuchia.

Apart from primitive and proto-lacertilian forms, two other small groups of extinct reptiles may well be included in the Eosuchia. *Champsosaurus* (Parks 262; Russell 263) is a long-snouted amphibious diapsid reptile from the late Cretaceous and earliest Tertiary. Since in earlier days the Rhynchocephalia were the only diapsids which were non-archosaurians, it was formerly placed in that order, and it is still placed there by both Hoffstetter (1955*b*) in the French Traité, and by Tatarinov. Retention of this disposition of *Champsosaurus* can be the result only of inertia. As an offshoot of the basic eosuchian stock, the Rhynchocephalia are characterized mainly by their specialized acrodont dentition; *Champsosaurus* retains a more primitive type of tooth implantation and is obviously a late surviving branch of the Eosuchia.

More than half a century ago, Merriam (260) described as *Thalattosaurus* a rather large aquatic reptile from the Triassic of California. The remains were imperfect and even the nature of the temporal region was in doubt. Merriam coined for this form the new group term Thalattosauria. Where the thalattosaurs should be placed in a scheme of reptile classification was long in doubt. Hoffstetter in 1955 (1955*b*) appended them to the Rhynchocephalia; Tatarinov (1964) includes them in the Lacertilia. The situation was clarified with the discovery in the Middle Triassic of Switzerland of a complete skeleton of *Askeptosaurus* (Kuhn-Schnyder 261), a form previously known only from fragmentary materials. This is a much smaller animal than *Thalattosaurus*, but one definitely related and with, fortunately, a well-preserved skull. The animal is a diapsid derivative, with its upper temporal opening reduced to a slit, and with a cheek region comparable to that of *Prolacerta*—the lower temporal bar partially reduced. It seems quite clear that we are dealing with an eosuchian offshoot.

Early lizards.—Perhaps overconservatively, I have classed lizards and snakes as two suborders of the order Squamata; many recent workers, including Tatarinov (1964), following the general trend to upgrade the status of groups of any sort, have not unreasonably elevated the Lacertilia and Ophidia (or Serpentes) to ordinal rank. Hoffstetter (1955*a*) followed the same course, giving lizards and snakes ordinal rank, and also raising the amphisbaenids to ordinal status; later, however, in an able review of the classification of the Squamata (274), he reversed this elevation and reduced lizards, amphisbaenids, and snakes to suborders.

Lacertilian diversification.—In my *Osteology of the Reptiles* of 1956 (225) I divided the "modern" lizard types into five infraorders: Iguania, Nyctisauria (Gekkota), Leptoglossa (Scincomorpha), Diploglossa (Anguimorpha), and Annulata (Amphisbaenia), with the Diploglossa subdivided into Anguoidea and Varanoidea (Platynota). Hoffstetter (1955*a*) used in the main a similar pattern, except that he followed Camp (1923) in combining my first two groups as a suborder Ascalabota and the next two as the suborder Autarchoglossa while, as noted above, he separated the amphisbaenids from the lizards as a distinct order in 1955 and placed the chamaeleons, as Rhiptoglossa, in an infraorder separate from the Iguania. Tatarinov (1964) followed much the same pattern as I did, but like Hoffstetter, separated the chamaeleons from iguanians and placed *Heloderma* in a discrete superfamily of the Anguimorpha.

The evolutionary story of the Lacertilia is divided into two very unequal portions, a Triassic-Jurassic chapter very poorly known, and a fairly well-documented story beginning in the late Cretaceous and extending to the present; between the two there is an intermezzo formed by the appearance of the mosasaurs and other aquatic lizards of the Cretaceous. Until recently our knowledge of pre-Cretaceous lacertilians was extremely poor and limited to a few forms which were inadequately known and of doubtful nature. One got the definite impression that the Lacertilia were a group which were late in development and did not blossom out until the Cretaceous. But it was, even so, certain that a pre-Cretaceous history had existed, since in the Cretaceous there are representatives of nearly all major lizard groups.

In recent years, however, the perspective has broadened considerably. Even as regards the Triassic, our gradually increasing knowledge of smaller reptile types now shows definitely that although certain forms are of dubious nature and others may be considered as eosuchians, the true lacertilian stage had been reached by the end of Triassic times. More than that, not only had true lizards evolved by the late Triassic, but they had already had time to radiate, for, curiously, the best-known certain lacertilians of the time are *Kuehneosaurus* (Robinson 264) and *Icarosaurus* (Colbert 1966), highly specialized in the development of rib-supported gliding planes.

The paucity of fossiliferous continental deposits has been responsible for our lack of adequate knowledge of lizard evolution in the Jurassic. Hoffstetter (1964) has recently ably reviewed our knowledge of the known faunas (cf. Cocude-Michel 1960, 1963). Up to the last two decades, only four genera of lizards had been described for the Jurassic—*Ardeosaurus*, *Euposaurus*, *Macellodus*, and *Yabeinosaurus*. Of these, it seems generally agreed that *Ardeosaurus* and *Yabeinosaurus* are members of the general nyctisaur group. *Euposaurus* was considered by Hoffstetter (1955*a*) and by Tatarinov (1964) to be of anguoid relationship; I have placed it tentatively in the Iguania, and Hoffstetter in his recent review believes it an agamid relative. Hoffstetter earlier considered *Macellodus* to be of uncertain position, while Tatarinov and I think it related to the Cordylidae, as does Hoffstetter himself in his 1964 paper. In 1947 *Teilhardosaurus* was described by Shikama from eastern Asia; it is dubious as to nature, although I have tentatively arrayed it with the gekko group. *Protaigialosaurus* (Kuhn 1958) is very probably, as its describer believed, a forerunner of the aquatic aigialosaurids of the Cretaceous. Hoffstetter in 1953 described *Bavarisaurus* and "*Broilisaurus*" —a homonym, renamed *Eichstattosaurus* by Kuhn (1958). The former was believed to be an iguanian, but Hoffstetter now concludes that it is a member of the gekko group; the latter is a gekko relative as well and close to, if not identical with, *Ardeosaurus*. *Changisaurus* (Young 1959*b*), described as a lizard, is apparently a chelonian (Baird 1964*a*). *Palaeolacerta* (Cocude-Michel 1960) was thought by its describer to be a lacertid, but Hoffstetter is currently doubtful about its relationship.

The Jurassic still fails to give us any broad picture of early evolution; it does show, however, that there was already considerable diversification, and most especially, the presence of several nyctisaurians (gekkotans) tends to confirm the belief of some earlier workers on lizard history in the antiquity and possible phylogenetic importance of the gekko group (although not of the obviously rather specialized "modern" family Gekkonidae) (Underwood 1954).

Post-Jurassic lizards.—As far as I am aware, little of importance has been published recently on the aquatic varanoid families of the Cretaceous. A new work on mosasaurs by Dale A. Russell is in progress, however.

From the late Cretaceous onward, continental deposits have yielded a relative abundance of lizard remains. In earlier years Gilmore was active in this field (particularly as regards the early Cenozoic forms); currently younger workers such as Estes (269), Etheridge, Hecht, and Tihen are active in this area in North America. In Europe, Hoffstetter is a major current worker, having published a number of valuable papers in the 1940's and having actively resumed work (despite recent distractions by fossil South American mammals) since his return from Eucador (274). Beyond the Cretaceous, particularly, we are mainly dealing with the early phases of the history of the existing families, and we are beginning to gain a moderately broad view of Tertiary lizard evolution.

Lizard-snake transition.—In Cretaceous and early Tertiary deposits are a number of forms, known mainly from fragmentary materials, that show vertebral structures similar, on the one hand, to those of varanoids and, on the other, to ophidians. Often grouped, in part or *in toto*, as "cholophidians," this series includes such genera as *Pachyophis* and *Simoliophis*. Where to place them in the system is doubtful; I have ranged them all, as advanced lizards, among the varanoids; probably more properly, Hoffstetter in 1955 placed them in the Ophidia as does Chudinov (1964). In my *Osteology* I associated *Palaeophis* and *Pterosphenus* with this same group and consequently classed them also as advanced varanoids; both Hoffstetter (1955*a*) and Tatarinov (1964), however, definitely include the *Palaeophis* group in the Ophidia as boid relatives.

Snakes.—In the *Osteology* I departed from orthodox classification of the snakes, for which I was taken gently to task by Dowling (1959) on such points as my "lumping" of the minor families of burrowers related to the boids, and so forth. Most of the points concerned do not affect the fossil story to any great extent, however. It is now generally agreed (Bellairs 273) that the Ophidia originated from forms more or less subterranean in habits, and small reptiles of this type are rare in the fossil record. Whether, for example, the Typhlopidae and

Leptotyphlopidae are related to one another, and whether either or both are only parallel developments from lizard ancestors rather than ophidians are moot points, for there is (with one exception) almost no trace of these families in fossil form. Underwood (1967) argues strongly for the ophidian nature of these two families.

The greater part of known fossil snakes are either members of the Boidae, which are the principal component of the superfamily Booidea or Henophidia, or are definitely members of the superfamily Colubroidea (Caenophidia) and assignable to the "advanced" families Colubridae, Elapidae, Hydrophiidae (sometimes combined with the elapids), and Viperidae. Additions to our knowledge of these typical snake assemblages are being made from time to time (for example, Auffenberg 277), but it is unusual that associated material is discovered, and well-preserved specimens of the delicate ophidian skull are particularly rare.

Problematical, however, in addition to the seemingly transitional forms discussed above, are the nature and position of certain other early forms, such as *Dinilysia* of the Upper Cretaceous of South America, *Coniophis* of the Cretaceous and early Tertiary of North America, and *Archaeophis* and *Anomalophis* of the Monte Bolca Eocene. *Dinilysia*, as the name suggests, was thought by its discoverer, Smith Woodward, to be allied to the South American boid burrower *Anilius* [*Ilysia*]; Tatarinov retains it in the Aniliidae, but there seems to be no close relationship here; I have placed it in a separate family close to the Boidae, and Hoffstetter inclines to include it, with doubt, in the Boidae. This form is now being studied by Estes (1967). Hoffstetter (1955*a*) placed *Coniophis* in a discrete family close to the boids, and I left it *incertae sedis* in 1956; however, Hecht (1959) gives convincing evidence for placing it in the Aniliidae. McDowell and Bogert (1954) suggested that the extremely elongate *Archaeophis* was an eel rather than a snake, but the argument does not seem well founded (Auffenberg 1959; Hoffstetter 274). Its position, however, is none too certain. Hoffstetter (1955*a*) placed it in a separate family among the "Cholophidia," and Chudinov (1964) considers it a member of the Palaeophidae; I have followed a suggestion of Hoffstetter's (274) in considering that *Archaeophis* and *Anomalophis* (Auffenberg 1959), from the same formation, are forerunners of the advanced colubroid group.

Rhynchocephalia.—When, well back in the last century, the unique nature of *Sphenodon* as a relatively unspecialized diapsid reptile was realized, high importance was attached to the Rhynchocephalia on the assumption that this order included the basic stock of all the Reptilia. Early in the present century their influence declined, and it presently became apparent that although the rhynchocephalians were relatively primitive in many ways, they were, in general, merely a conservative side branch of the early lepidosaurians with the development of acrodonty and of a beak as distinctive characters. A further degrading is done by Tatarinov (1964) in including them in the order Eosuchia.

The "main line" of rhynchocephalian evolution is poorly documented, but the data we have indicate that *Sphenodon*-like forms had come into existence in the early Triassic and were widespread, if poorly represented in the fossil record, until the Jurassic. From this time up to the Recent the record is barren.

A surprising development of recent decades has to do with the rhynchosaurs of the Triassic—relatively large types with parrot beaks and a highly specialized dentition. A few specimens of these forms have been known for a century, since the days of Huxley and Owen, in Great Britain and India, and a few primitive forms were presently discovered in South Africa. They long appeared, however, to have been of little importance. But they now seem to have been an exceedingly important element in the Middle Triassic fauna, in the southern continents at least (Huene 278, 279, 280). I have on several occasions summarized this history (Romer 1962*c*, etc.). They are common in the Middle Triassic Manda beds of East Africa and, in the roughly contemporary Santa Maria and Ischigualasto beds in South America, they are the most abundant forms present. Whether or not they were also abundant in the northern continents at that time we do not know, owing to the almost complete absence of Middle Triassic continental sediments. There is, of course, no reason to believe that they were suddenly "extinguished" at the end of the Middle Triassic, for while they are unreported in the Keuper of continental Europe, specimens in Keuper (or presumed Keuper) beds have long been known in

England. The same is true of beds in Elgin, Scotland, and in India which are presumed to be of late Triassic age, and they are now reported from Nova Scotia (Baird and Take 1959) in beds thought to represent a low level in the Upper Triassic Newark series.

Miscellaneous lepidosaurs.—Apart from various forms from the Triassic and Jurassic which can be rather definitely assigned to one of the three established lepidosaur orders, there remain a number of types of problematical relationship. In the Ladinian deposits of southern Switzerland, worked by Peyer and now being continued by Kuhn-Schnyder, a number of odd types unknown (or almost unknown) elsewhere have been difficult to place, mainly because of incomplete knowledge of cranial structure. *Macrocnemus* (Peyer 356, part 12) now can be definitely placed as just about astride the eosuchian-lacertilian boundary. *Clarazia* and *Hescheleria* (Peyer 356, parts 10 and 11), very incompletely known, may be appended with doubt to the Rhynchocephalia, as I and other recent writers have done. The grotesque long-necked *Tanystropheus* (Peyer 356, part 2) is problematical in great measure because of incomplete knowledge of temporal structure. I have argued (Romer 1947) that this weird shore-dweller is part of an early euryapsid radiation; others, very probably correctly, have maintained that when and if the temporal structure is ascertained, it will be found to be lepidosaurian, and Peyer and Kuhn-Schnyder (1955) in the French Traité, and Rozhdestvenskii and Tatarinov (1964) have allied it to the Lacertilia.

Two forms from the lithographic limestones are a perennial puzzle: *Pleurosaurus* [*Acrosaurus*] (Huene 1952*b*) and *Sapheosaurus* [*Sauranodon*]. They are rather certainly lepidosaurians, but their placement in the group is dubious. In the second edition of the *Vertebrate Paleontology* I placed, with doubt, *Acrosaurus* in the Eosuchia and *Sapheosaurus* in Rhynchocephalia. In the present third edition I have placed both in the latter order; in the Piveteau Traité, Hoffstetter has done the same, but Rozhdestvenskii and Tatarinov, while placing *Sapheosaurus* in the Rhynchocephalia, have shifted *Pleurosaurus* to the order Araeoscelidia in the euryapsid group. As far as I am aware, there has not been any recent original work of note on these forms.

11

Ruling Reptiles

Archosaur origins.—It was long held by most workers that the two-arched reptiles, the Diapsida, were a unit, a monophyletic group. But with the realization that the Squamata were of "diapsid" origin and sprang from a presumably common eosuchian ancestry with the Rhynchocephalia, it became apparent that the two-arched reptiles and their derivatives included two very diverse groups, the Lepidosauria (essentially conservative, except for the snakes) and, quite in contrast, the great group of forms of which the crocodilians, pterosaurs, and dinosaurs were members, and from which sprang the birds as well. For this second radiation Huene's appropriate term Archosauria was available. But despite this evidence of two different types of evolutionary radiation among two-arched reptiles, there has tended, conservatively, to be a belief that basically lepidosaurs and archosaurs were related; this belief has been sometimes expressed by retention of a subclass Diapsida, subdivided into archosaurian and lepidosaurian groups. I have long doubted whether there ever existed a common diapsid ancestor of the two, and in the current state of our knowledge this appears highly improbable. Utilizing readily recognizable key characters, we have seen that the earliest and most primitive known lepidosaurs had already developed a short jaw and an incipient otic notch, and if the millerettids are truly ancestral to the lepidosaurs, this type of structure had already come into existence in a stage before the development of temporal openings. Although the ancestry of the archosaurs is incompletely known, we find that in certain forms (such as *Chasmatosaurus*) which are definitely archosaurian, the jaws still extended well back of the plane of the occiput in primitive fashion, and there was no otic notch. It thus appears almost certain that there was no common ancestor of the two groups which was diapsid in nature, and no evidence whatever that lepidosaurs and archosaurs are at all related except that, of course, both are descended from the basic cotylosaurian reptilian stock. Although even the most primitive of Triassic archosaurs had already developed a number of features characteristic of this great group, there is nothing in the structure of these forms which negates the probability that they have descended, rather directly and independently, from captorhinomorph ancestors.

Bipedality vs. quadrupedality.—Primitive reptiles were, of course, quadrupedal, and various archosaurs, living and extinct, are certainly quadrupedal types—for example, phytosaurs, crocodilians, sauropod dinosaurs, stegosaurs, ankylosaurs, and ceratopsians. But there was, almost unique in vertebrates, an obviously strong trend towards bipedal locomotion in the group. The theropod dinosaurs among the Saurischia were definitely bipeds; the ornithopods among the Ornithischia were at least partially bipedal; obviously both birds and pterosaurs were descendants of bipedal archosaurs in which the "arms" had been freed from locomotor function. Many of the most striking skeletal adaptations of most archosaurs seem associated with bipedality: large stout hind legs posed close into the body, front legs much smaller and less well adapted; a specialized pelvis apparently associated with bipedal posture; a powerful tail, useful as a balancer in upright pose. Even in many forms of archosaurs which are quadrupedal, these characters are apparent, as witness the great disparity in limbs in stegosaurs, ceratopsian dinosaurs, and nearly all sauropods.

In consequence, the thesis (to which I have long subscribed and from which Gilmore was one of the few dissenters) was formulated that the bipedal trend came early in archosaur evolution, and that most later archosaurs which were quadrupedal were actually descended from earlier bipedal ancestors.

Currently, however, there is a reaction against this theory. Much of the counterargument is not as yet published but has been discussed by Attridge, Crompton, and Charig, and I have discussed it with the last named. This reaction began in great measure through the discovery that in *Melanorosaurus* of the late Middle Triassic of South Africa, we are dealing with a form closely related to the ancestry of the sauropods but which was at that time definitely quadrupedal. There had hardly been time, the argument runs, by the Middle Triassic, for the development of bipedalism and then a secondary slump to all fours. Might not the sauropod ancestors have remained quadrupedal throughout? And if sauropods are primitively quadrupedal, might not the same be true of other quadrupedal archosaurs which are generally regarded as having been of bipedal ancestry? May not the bipedal trend have been the exception rather than the rule amongst archosaurs?

But if quadrupedality was the prevalent mode, why the pronounced specializations in hind legs, pelvis, and tail, developed almost universally in the group? A suggested answer is that they were useful generally as improvements in quadrupedal locomotion. How did they originate? Possibly as adaptations for an amphibious mode of life. Such archosaurs as the crocodilians in late times and the early phytosaurs of the Triassic were amphibious in habits; some of the dinosaurs, notably the sauropods, show amphibious trends. Possibly the ancestral archosaurs were water loving (not impossibly retaining this habitus from ancestral reptiles?), and a powerful tail and hind legs (cf. *Mesosaurus*) can be useful swimming adaptations. Once developed, these characteristics could prove useful in terrestrial quadrupedal locomotion and especially useful in such forms as did become bipeds.

This is an interesting argument, deserves careful consideration, and certainly has a considerable amount of justification. It has been generally assumed that the bipedal trend was an early and almost universal one among the ancestral archosaurs, the thecodonts. But as our knowledge of this order gradually increases, it becomes more and more apparent that there existed a considerable variety of thecodonts, which were rather surely primitive quadrupeds, in whose pedigree no prior bipedal stage could well have existed. Notably the Proterosuchia of the Lower Triassic; despite the fact that in such a form as *Erythrosuchus* the archosaur locomotor specializations were already present in at least an incipient stage, this large and cumbersome beast was definitely a quadruped, without any implication of bipedalism.

One should not, however, in enthusiasm press this argument too far, and it seems probable that the actual situation lay between the two extremes of theory. It seems certain that at least a fraction of thecodonts never tended at all towards bipedality, and quite possibly certain quadrupedal archosaurs which have been suspected of having had at least some bipedal trends in their ancestry and then reverted may prove innocent. But any sweeping claim that all quadrupeds in the group are primitively quadrupeds is improbable. For example, in the ornithischian dinosaurs, the presumably primitive ornithopods are partially to almost completely bipedal, but the other subdivisions of the group—stegosaurs, ankylosaurs, ceratopsians—are quadrupeds. Are they primitively quadrupedal? The high disproportion between front and hind legs in ceratopsians and to an absurd degree in *Stegosaurus* had argued strongly for reversion in these cases. Ornithischian early history is very incompletely known, and there is no real evidence for stegosaur ancestry. But in the case of the ceratopsians, reversion seems clear. *Psittacosaurus* and *Protiguanodon* of Mongolia, as mentioned later, are bipeds but generally are admitted to lie on the line of the quadrupedal ceratopsians.

Some archosaurs, then, were certainly primitive quadrupeds. In some cases it is as yet uncertain whether quadrupedality is primary or secondary. But in at least one instance reversion seems clear. And the fact that bipeds evolved in both dinosaur orders as well as surely in bird and pterosaur ancestors does indicate a strong bipedal trend within the early archosaur stock.

Thecodonts.—During the Triassic—and the Triassic alone, as far as we know—there existed a considerable variety of forms which are definitely archosaurian in nature but not members of advanced groups such as dinosaurs, crocodilians, or pterosaurs. In earlier days they were often classed as Parasuchia, including only the amphibious phytosaurs, and Pseudosuchia, a rather inappropriate term, under which were placed all other primitive members of the subclass. Currently it is more general to place them all in a common order Thecodontia, following Watson (1917) who revived an old Owen term. Huene (226), Hoffstetter in the Piveteau Traité (1955*c*), and Rozhdestvenskii and Tatarinov (1964) in the recent Russian Treatise, use Pseudosuchia and Parasuchia as suborders, although Hoffstetter divides the Pseudosuchia into superfamilies somewhat comparable to the three suborders into which (as discussed below) I have divided the erstwhile pseudosuchians. But how to sort out the phyletic relationships among the various thecodonts and treat them in systematic fashion is problematical. Our knowledge of the group is gradually increasing, and it is to be hoped that specialists in the group may presently work out some reasonable scheme of classification. Reig (1961) has recently discussed this situation and has in preparation a thorough review of the proterosuchians and early archosaur history. Tentatively, I have arrayed the thecodonts in four suborders—Proterosuchia, for primitive quadrupeds, with little or no armor; Aetosauria, for armored quadrupeds such as *Aetosaurus* and *Typothorax*; Phytosauria (or Parasuchia) for the phytosaurs; Pseudosuchia, for "progressive" forms with bipedal tendencies.

In the Proterosuchia the best-known forms have been *Chasmatosaurus* and *Erythrosuchus* from the early Triassic of South Africa; the same or comparable forms have in recent years been discovered in northern continents, in China and Russia (notably *Shansisuchus*, Young 1964). Both show definitely primitive characters in palatal construction, lack of development of the archosaur type of otic notch and so forth (Hughes 281). But both long-snouted *Chasmatosaurus* and large and awkwardy built *Erythrosuchus* are obviously specialized; they are not in themselves truly ancestral archosaurs, but early side branches of a primitive archosaur stock which is unknown, but must have been developing in late Permian times, since more advanced forms of thecodonts, such as *Euparkeria*, were already developed in the Lower Triassic. With the Proterosuchia I have placed (in my text) some further Lower and Middle Triassic forms, mainly genera sometimes included in a family Rauisuchidae (Reig 1961) which are incompletely known but seemingly rather primitive in nature. This is probably incorrect; these forms appear to be rather more advanced than the proterosuchians.

As a second group I have erected a suborder Aetosauria for a series of highly armored quadrupeds such as *Aetosaurus* of the German Keuper, *Stagonolepis* (Walker 283) of the Elgin Trias, and *Typothorax* (Sawin 287) of the western American redbeds. They are mainly known from the Upper Triassic (indeed, the similarity between *Stagonolepis* and *Aetosaurus* is a principal argument for claiming the Elgin Triassic to be late Triassic in date) but, as *Typothorax* witnesses, were rather varied in nature, suggesting that they had been in existence for some time; that this is the case is demonstrated by the presence of aetosauroids in the Middle Triassic of Argentina (Casamiquela 1961). In the case of the aetosaurs (and the phytosaurs), proponents of a strongly bipedal thecodont trend have suggested that the ancestors had been somewhat bipedal before assuming their obviously quadrupedal (and probably rather amphibious) mode of locomotion; but in default of proof to the contrary, primitive quadrupedality is possible, despite considerable disparity in leg length.

Despite reasonable protests against too great emphasis on bipedal trends in archosaurs, it is evident that such tendencies were strongly present among the thecodonts, and that this trend began early in thecodont history is evinced by *Euparkeria* of the early Triassic of South Africa, in some regards a rather primitive thecodont, recently redescribed in excellent fashion by Ewer (282). I have restricted the term Pseudosuchia to thecodonts tending to bipedalism. There are a number of forms of this sort in the late Triassic, and others have come to light in recent decades in the Middle Triassic of Africa and South America. Some presumably represent sterile end lines, but did we know the smaller Triassic thecodonts better, we should expect to identify lines leading toward pterosaurs and birds. Unfortunately, we have as yet no clues in this regard. Nor do we as yet know any thecodonts suggestive of ornithischian dinosaur ancestry.

For the Saurischia, however, the situation is very different. It would take little modification to transform certain pseudosuchians into bipedal saurischians, and the boundary between the two orders is a vague one, of which it is hard to be certain unless the anatomy of a form concerned is completely known. Charig has named (but not as yet described in print) forms from the Middle Triassic of East Africa which appear to be just below the saurischian level, and only separable from that group by the fact that the acetabular socket is imperforate (a useful if minor diagnostic feature). *Ornithosuchus* of the Elgin has long been considered a typical pseudosuchian; Walker, however, in a recent restudy (322) considers it a saurischian, although a very primitive one; and I have included it in the Saurischia, although with considerable qualms. Very probably the saurischians arose in mildly polyphyletic fashion from two or several pseudosuchian forms.

The Phytosauria (Parasuchia) have long been known from the later Triassic of both Europe and North America and are also known from India. They are exclusively Upper Triassic, as far as known, with one exception—the specimen named *Mesorhinus*, which, although apparently a typical phytosaur, was said to come from the Bunter Lower Triassic of Germany. This is puzzling; possibly the group was in the process of developing in the unknown Middle Triassic of the northern continents. But the presence of a phytosaur in the Lower Triassic seems unlikely. As Gregory (290) has suggested, there may have been some mistake about the place of origin of the *Mesorhinus* specimen, and it may actually be from the red deposits of the Upper rather than the Lower Triassic. The taxonomy of the phytosaurs is in a confused state. There is little variation, apart from skull shape, among phytosaurs, yet about three dozen generic names are present in the literature. Not long ago Gregory (290) and Westphal (1963) almost simultaneously proposed revisions of phytosaur systematics, with a strong "boiling down" of genera. The two proposals differ to a considerable degree. I have used Gregory's interpretation, provisionally. But despite these attempts at simplification, the systematic situation is still far from clear (cf., for example, Colbert 1965, pp. 19–23).

Generally included among the thecodonts are a number of genera which to a variable degree suggest crocodilian ancestry. I have in my text transferred certain of them to the latter group, as noted later. Complicating the whole thecodont picture is the presence in many members of the order of the crocodilian type of intratarsal foot joint (Krebs 1963, 1965).

Crocodilians.—These reptiles are among the commonest of fossils in beds from the Jurassic onward, for their massively built skulls (in contrast to the feebler skull structure of many other groups) tend to survive fossilization in good style and, further, their amphibious habits place them, in life and in death, in an environment in which sediments which may preserve them as fossils are being laid down. From the Jurassic on, and in recent decades as in earlier times, there have been numerous reports of new finds of typical mesosuchians and eusuchians, including description of a dozen new genera (notably from China, Young 1948*b*).

The general scheme of arrangement of the mesosuchian and eusuchian crocodilians has for the most part reached a stable stage. Kälin (1955) compares the classification he used with that of early works, and I have (mentally) continued this table by comparing the Mook and Kälin classifications with those of Huene (226) and myself (225), and of Rozhdestvenskii and Tatarinov in the recent Russian Treatise (1964). Six mesosuchian families are recognized by all writers—Metriorhynchidae, Teleosauridae, Pholidosauridae, Atoposauridae, Notosuchidae, Goniopholidae—the only variant here being that, following many early workers, I considered the specialized marine Metriorhynchidae to constitute a separate suborder Thalattosuchia. By Kälin and the Russian writers the Dyrosauridae, which Huene and I placed in a separate family, are lumped with the Pholidosauridae, whereas on the other hand, *Libycosuchus* is split off from the Notosuchidae to become the type of a separate family; the Russian authors alone separate *Bernissartia* from *Hylaeochampsa*. Among the Eusuchia, all are agreed that *Hylaeochampsa* represents a distinct primitive family, and that the Stomatosuchidae are distinctive. There is, however, variation in the treatment of alligators, crocodiles, and gavials. Mook and Huene place alligators and crocodiles in separate families; Kälin, the Russians, and I united them in the Crocodylidae, while I, in contrast to the others, included the gavials in the Crocodylidae as a subfamily.

An interesting recent development in the crocodilian story has to do with the Sebecosuchia, a highly predaceous subgroup of crocodilians. The story began with the discovery of *Sebecus* in the Eocene of South America by Simpson (Colbert 296). Later, Price (1950) described another related carnivore, *Baurusuchus* from the Cretaceous; still later, he described (1955) a further Upper Cretaceous form, *Peirosaurus*; Langston (1965*b*) has reported *Sebecus* from the Miocene and suggested (1956) that this group might have been worldwide, rather than confined to South America. Not improbably this is the case; for example, Berg (1966) reports a *Sebecus*-like form from the European Eocene.

Obviously, the crocodilians are of thecodont descent, and crocodilian similarities have been noted in several instances. *Protosuchus*, from the Triassic-Jurassic boundary in Arizona (Colbert and Mook 295), has definite crocodilian characteristics and is frequently included in the order Crocodilia; *Sphenosuchus* of the South African Triassic is also suggestive of crocodilian relationship. Somewhat closer, however, are such forms as *Notochampsa* of South Africa and *Proterochampsa*, recently discovered in Argentina (Reig 1959). I have, for the time being, after discussion with William D. Sill (who published a study of the last-named form in 1967) arrayed these four seeming crocodile relatives in two suborders, the first two as Protosuchia, the last two as Archaeosuchia.

Pterosaurs.—For the most part, we know no more about these flying reptiles than we did several decades ago. Items of recent interest are the description by Riabinin (1948) of a new rhamphorhynchoid, *Batrachognathus*, from western Asia, and, most especially, Young's description (304) of the slender-beaked *Dsungaripterus* from the Lower Cretaceous of Sinkiang. Brink (1963), in parallel with his argument for the erection of a separate class for the presumably warm-blooded Therapsida, has proposed that separate class status be given to the Pterosauria, which may have been warm-blooded as well; in this he is followed by Kuhn (1967). The proposal is not illogical, but I see no advantage to be gained by this fragmentation of the Reptilia.

12

Dinosaurs

During the early decades of this century a vast amount of interesting and valuable work was done on dinosaurs. Although less in amount and importance, considerable new discovery and description has been accomplished in recent years. Central and eastern Asia have been most productive. Exploration by Russians of the Mongolian beds, which furnished specimens for the American Museum expeditions, has yielded a great amount of good material (Maleev 1952, 1954, 1955*a*; Rozhdestvenskii 1952*a*, etc.), and work in these beds is being continued by Polish expeditions. Much material has been obtained in China from all three Mesozoic periods, described principally by Young (1947*b*, 1948*a*, 1958*a–d*, 1959*a*; also, Hu 1964). Bohlin (1953) has described an important Cretaceous fauna from Kansu. Lapparent has described Upper Cretaceous forms from southern France (1947), Morocco (1955), Portugal (with Zbyszewski 1957), and the Central Sahara (1960). A dinosaur-egg locality in Provence, first discovered a century ago, has been profitably reworked (Dughi and Sirugue 1957). Among recent scattered North American finds are a new Cretaceous sauropod (Gilmore and Stewart 1945), a new *Gorgosaurus* skull (Gilmore 1946*a*), an Upper Cretaceous fauna from Utah (Gilmore 1946*b*), hadrosaur material from New Jersey (Colbert 1948), a Lower Cretaceous theropod from Oklahoma (Stovall and Langston 1950), a new Canadian hadrosaur (Sternberg 1953), a hadrosaur from Alabama (Langston 1960), an ankylosaur from Kansas (Eaton 1960), and good material of the Triassic coelurosaur *Coelophysis* (Colbert 1964). Publication was continued by Janensch (1955, etc.) on the Tendaguru collection in Berlin. In South America a few remains have been described from Patagonia (Cabrera 1947; Casamiquela 1964), and Triassic specimens have been found in Brazil and Ischigualasto, Argentina

(Huene 238; Reig 1963). Colbert, Cowles, and Bogert (1946) have considered temperature problems in dinosaur life—and death; Colbert (1951) has discussed environmental factors in general and published (309) on probable weights in dinosaurs.

Colbert has published popular accounts of dinosaurs in 1951 and 1961 (305) and general accounts have been given, almost simultaneously, by Lapparent and Lavocat in the Piveteau Traité (1955), by Huene in his work on lower vertebrates (226), and by me in my *Osteology of the Reptiles* (225); more recent is a general account by Rozhdestvenskii and Tatarinov (1964) in the Russian Treatise. For the most part there is little difference in the four treatments; on the whole a rather monotonous uniformity. Lapparent and Lavocat use the terms "Sauripelviens" and "Avipelviens" rather than Saurischia and Ornithischia. But while their terms are more properly descriptive (after all, it is the pubis, not the ischium that is diagnostic), the Seeley terms are so firmly established in the literature that it seems inadvisable to change.

For nearly two-thirds of a century Huene has been the leading student of the Saurischia, and it is hence not unexpected that all three accounts follow his lead in dividing the bipedal members of the order into three groups—the lightly built Coelurosauria, the Carnosauria, including the giant carnivores, and the more clumsily built Prosauropoda of the Triassic, ancestral to the Sauropoda. The arrangement, however, differs in the three works. Lapparent and Lavocat follow a customary procedure in placing all three bipedal assemblages in a single suborder Theropoda, although recognizing the descent of the sauropods from prosauropod ancestors. Huene adopts a different course, with two suborders, in which the Coelurosauria are separated from all other saurischians, lumped as the Pachypodosauria. This suborder is divided into Carnosauria, Prosauropoda, and Sauropoda. This expresses his view that the primary cleavage in the Triassic was into the lightly built coelurosaurs and more ponderous bipeds, among which occurred a split into carnosaurs and prosauropods.

A notable contribution to saurischian history is Walker's study of *Ornithosuchus* (322). This form has been generally considered as a rather generalized bipedal thecodont, possibly ancestral in a broad way to saurischians and possibly other advanced archosaurs as well. Walker claims, however, that *Ornithosuchus* is actually a carnosaur. But if so, it is certainly a carnosaur of very primitive aspect, with such archaic features as the retention of dorsal dermal armor. Certainly it is a form not far over the ordinal boundary from the Thecodontia, and I do not feel at all sure that consideration of it as an advanced, near carnosaur thecodont would not be preferable.

It is not improbable that the Saurischia had a di- or polyphyletic origin, and that coelurosaurs came from a discrete thecodont stock. If this be the case, there is justification for Huene's classification as a closer approach to a "natural" phylogenetic one than that adopted by Lapparent and Lavocat and by me. Huene differs in that *Ceratosaurus* is considered to be a giant coelurosaur rather than a carnosaur. The little carnivores *Velociraptor* and *Saurornithoides* from the Mongolian Upper Cretaceous are included by Huene and me in the coelurosaurs; Lapparent and Lavocat, however, included them in the Carnosauria, with the statement (not amplified) that *Velociraptor* is "typiquement un Mégalosauridé."

I had at first planned to use the customary theropod-sauropod cleavage in my text, but was influenced by two works—Colbert (315), and Charig, Attridge, and Crompton (315*a*), which appeared almost simultaneously in 1964–65. Both papers discuss the large, mainly bipedal, dinosaurs of the late Triassic and, while differing somewhat in treatment, point out that none of these, whether apparently herbivorous or carnivorous, have any relationship to later theropods, but are related to sauropod ancestry. The second of these two papers calls particular attention to *Melanorosaurus* of South Africa, which appears to have progressed far toward sauropod conditions well before the end of the Triassic. Colbert suggests using the term Palaeopoda for the Triassic forms here considered; I have, however, followed Charig *et al.* in conservatively using Prosauropoda for them and bracketing them with the sauropods in a suborder Sauropodomorpha.

A proper classification of the great amphibious sauropods has been the despair of everyone working on the group including, I strongly suspect, even Huene, despite his long study of them. The reasons for our difficulties are apparent. Few complete skeletons exist; feet and skulls are rare; many of the numerous described forms are based on fragmentary material. In Huene's classic work of 1932 (316), he ranged them in two families, Cetiosauridae and Brachiosauridae, each with four subfamilies. In 1956 (226) he followed a suggestion of Janensch in ranging them in two "family groups" (for which he unfortunately uses improper family endings). These are the "Bothrosauropodidae," including Cetiosauridae, Astrodontidae, Brachiosauridae, and Camarasauridae, and the "Homalosauropodidae" with the families Titanosauridae, Apatosauridae, Diplodocidae, and Dicraeosauridae. The two groups are defined on cranial characters, the first having large narial openings placed in front of the eyes, and spoon-shaped teeth; in the second, the relatively small nostrils are above the orbits, and the teeth simple pencil-shaped structures. As far as I am conversant with the material, this cleavage is reasonable for the forms in which the skulls are known; but they form only a small percentage of known genera. It will be noted that this classification is very different from Huene's in 1932, for the forms which he then used as types of his two families are both included here in the "Homalosauropodidae." Lapparent and Lavocat array the material in six families—Cetiosauridae, Brachiosauridae, Camarasauridae, Astrodontidae, Diplodocidae, and Titanosauridae—an arrangement that agrees fairly well with Huene's, except that they include his apatosaurs and dicraeosaurs in the Camarasauridae, and that a number of the more poorly known genera are variably assigned in the two cases. In my own work of 1956 and in my new edition I adhere to a two-family arrangement, my Titanosauridae having much the composition of Huene's "Bothrosauropodidae" and my Brachiosauridae being essentially his "Homalosauropodidae." It will be a long time, if ever, before we obtain a valid, comprehensive picture of sauropod classification and phylogeny.

Ornithischians.—Studies on the evolution and classification of this great group have been severely handicapped by a dearth of material earlier than the late Jurassic. For the earlier part of the Jurassic we have had to rely on such few forms as chanced to be washed into the sea, since most Jurassic deposits are marine. Among the Saurischia, late Triassic forms are common, and the transition from thecodonts to dinosaurs is readily visible.

In the Ornithischia the story is very different. No known thecodont is suggestive of ornithischian ancestry. That ornithischians were actually relatively rare is indicated, for example, by the paucity of tracks possibly made by ornithischians among the wealth of footprints of the Connecticut Valley (Lull 317; Colbert 1963). Until recently we knew only a single jaw fragment (*Geranosaurus*) which seemed to be remains of a Triassic ornithischian (Huene 1950). (A supposed ornithischian Triassic pelvis, *Poposaurus*, is considered by Colbert [1961] to be more probably saurischian.)

In the last few years, however, the situation has begun to improve. Crompton and Charig (334) have described as *Heterodontosaurus* an ornithischian skull (of rather specialized nature) from the South African late Triassic redbeds; Crompton, Charig, and Attridge have very recently collected further ornithischian remains from South Africa; the fragmentary *Lycorhinus* jaw from the same deposits, once thought to be that of a therapsid, is now identified as ornithischian; incomplete materials from the Chinese Triassic have been described as *Tatisaurus* (Simmons 1965); an ornithischian jaw fragment has been discovered by Baird in the Triassic of Nova Scotia.

The fact that a variety of small ornithischians which appear to be of a very primitive nature are present in the early Cretaceous (as *Hypsilophodon*) or the late Cretaceous (as *Thescelosaurus*) suggests that the evolutionary story of the ornithischians may have been a relatively conservative one, with a central, relatively unchanging stock of primitive forms which persisted until the end of the Mesozoic, and with specialized lines branching off from it, to give the group phylogeny a treelike pattern, rather than the "bush" effect commonly seen or suspected. One small but significant indication that this may have been the case is the fact that premaxillary teeth, generally absent, are persistent in the late Cretaceous, not only in presumed conservative forms but in definitely aberrant ones such as the dome-headed *Stegoceras* and the "horned" dinosaur *Protoceratops*.

It is generally agreed that the bipedal members of the group be included in the suborder Ornithopoda, with a "main line" of evolution of which *Hypsilophodon* is representative of a primitive stage, *Camptosaurus* and *Iguanodon* of a somewhat more advanced one, and with the curious duckbills of the Upper Cretaceous, the hadrosaurs, the terminal group. Generally these forms are arrayed in three families, with the duckbills forming the family Hadrosauridae. Huene, however, appears to be so impressed by the variety of curious crests assumed by the hadrosaurs that he divides them up into five families; Lapparent and Lavocat satisfy themselves with a subfamilial arrangement. Attempts at subdivision of the group are dangerous. The known time of existence of the hadrosaurs was a relatively short one, geologically, including only a small fraction of the Upper Cretaceous, and little can be determined about phyletic lines. Ostrom (341, 1962) has made a valuable study of the hadrosaur skull and, considering their history, sees little evidence suggestive of evolutionary sequences of genera; he pictures (341, Fig. 78) their evolution as bushlike, nearly all forms arising independently from a common base.

With regard to the striking development of crests in hadrosaurs, Ostrom refutes, reasonably, various earlier suggestions about crest function, mainly related to presumed underwater feeding, but as a substitute can only offer the suggestion that the convoluted tube traversing the crests afforded additional area for olfactory epithelium. It is difficult to suggest a reason why some (but not all) members of the family should demand extraordinary olfactory powers.

It is curious how errors, once implanted in the literature, persist. A typical example is the frequently copied Dollo illustration of the *Iguanodon* pelvic girdle. In ornithopods generally a flange develops on the ischium which extends ventrally medial to the pubis. When the Bernissart "herd" of iguanodons was discovered, the ornithischian pelvis was little known, and in the first skeleton mounted in Brussels the ischia were apparently reversed and the flange consequently descended outside the pubis. Dollo figured it in this position in 1883 (Pl. III, Fig. 2). Skeletons mounted later had the ischia in proper position, but this point apparently escaped the attention of the usually meticulous Dollo, who as late as 1923 republished his original erroneous figure. But that was not the last of it; Lapparent and Lavocat (1955, p. 786), after the lapse of three-quarters of a century, repeat Dollo's error, and the same is true of the interesting recent account by Casier (339, Fig. 16C) of the Bernissart iguanodonts.

In the bipedal ornithopod assemblage there are two interesting aberrations from the main evolutionary line. One includes *Psittacosaurus* and the ineptly named *Protiguanodon*, from Asiatic beds once thought to be Lower Cretaceous. Their parrot-like beaks and other features strongly suggest that they are essentially ancestral to the ceratopsians of later stages in the Upper Cretaceous, and it seems to me not impossible that the bone in *Psittacosaurus* identified as the premaxilla is actually a rostral. Most extraordinary is the type of skull seen in *Pachycephalosaurus* and *Stegoceras*—the latter generally, but apparently incorrectly, termed *Troödon* (Sternberg 1945; Russell 1948). Here the skull is crowned with a massive dome-shaped structure of solid bone. With this skull is associated a rather lightly built bipedal skeleton. The association of skull and skeleton was at one time doubted, but now seems assured. The palatal structure of the *Stegoceras* skull is suggestive of the ankylosaurs, but there is obviously no direct genetic connection.

Bipedal locomotion seems never to have been as firmly established in ornithischians as in many of the saurischians, and the order includes a great array of armored and horned quadrupeds. In default of adequate evidence regarding early evolution of the order, it is possible to claim that part or all of these forms are primitively quadrupedal rather than having reverted from an earlier bipedal mode of life. Such a claim, however, cannot be made in the case of the ceratopsians, for, as noted above, it seems clear that they are descended from bipeds of the *Psittacosaurus-Protiguanodon* type. For *Stegosaurus* and its kin there is no proof, one way or the other, as to the pedigree; but I find it impossible to believe that the grotesque pose of these forms—the shoulder region very low, the pelvis highly elevated—can have come about as the result of a history of simple quadrupedality.

How to classify the quadrupeds has presented problems, particularly because of the dearth of Jurassic materials. Nopcsa (227) advocated assigning all the quadrupeds to a single group, Thyreophoroidea, but the horned dinosaurs are so clearly a discrete group that all later workers have agreed that they constitute a distinct suborder (or superfamily), the Ceratopsia (Ponderopoda of Huene). Opinions have differed over the classification of armored forms.

Stegosaurus and its close relatives of the late Jurassic and early Cretaceous have long been well known with their characteristic structures, such as the double row of plates and spines extending the length of the back and tail and a very distinctive pelvis, the pubis having both a good anterior process and a long, straight, broad posterior stem apposed to a similarly developed ischium. Later to be recognized were the members of a second, late Cretaceous series of armored forms, the ankylosaurians or nodosaurs. These were at first, when known mainly from assorted spines and plates, confused with the stegosaurs. It is unfortunate that even now, when adequate materials have been known for more than half a century, no adequate account of this important group has ever been given. Such a study was begun a number of years ago by Brown and Schlaikjer, but never completed. The ankylosaurians' distinctive characters were never realized until, some decades ago, when studying the pelvis in ornithischians, I had occasion to survey the splendid material in the American Museum of Natural History and found, to my astonishment, that not only in pelvic structure but in almost every other feature, these "reptilian tanks" were radically different from the stegosaurs and merited their separation as a suborder Ankylosauria (Romer 1927). Apart from the fact that the armor is a carapace rather than a mere pair of spine rows, the broad massive skull is very different from that of the stegosaurs, the body proportions are radically different from those of the relatively tall, slender stegosaurs with their very long hind legs, and the pelvic structure is unique. The ilium is rolled outward to a variable degree over the acetabulum, the ischium is downwardly curved, much as in ceratopsians; and most diagnostic, the pubis is strongly reduced, not only the posterior branch but, unique among ornithischians, the anterior process as well; there is typically left nothing more than a basal part forming a segment of the solidly built and (remarkably) unperforated acetabulum.

Despite this sharp contrast with stegosaurs, various authors, including Huene and Lapparent and Lavocat, have tended to retain the two groups in a single suborder or superfamily (Thyreophora, Huene; Stegosauroidés, Lapparent and Lavocat). For the most part the reason seems to be only the fact that both types are armored. This is meaningless. A trend for armor development is deeply ingrained in the Archosauria (as witness the numerous armored thecodonts and the Crocodilia), and was surely present in primitive ornithischians. It is possible that there may have been an early armored common ancestor of stegosaurs and ankylosaurs, but there is no reason to believe that there was, and no evidence for it. There is currently no reason whatever for uniting stegosaurs and ankylosaurs in a common group. Maleev (1952) has described as *Syrmosaurus* an Upper Cretaceous armored form from Asia, for which, on grounds which do not seem to me adequate, he has erected a new family. Lapparent and Lavocat have seized upon this form as giving a basis for fusing stegosaurs and ankylosaurs, stating that while it is ankylosaurian in most regards, its pelvis is stegosaurian. I do not understand this claim. The ilium, as figured by Maleev, is clearly ankylosaurian; the curved ischium is likewise ankylosaurian; and it is significant that the pubis, much reduced in typical ankylosaurs, is not identified.

I have separated from the typical ankylosaurs a few seemingly lightly armored forms as a separate family, the Acanthopholidae. This, however, is rather arbitrary; most members of this group (and various forms which I have placed in the Nodosauridae) are known only from fragmentary material.

At the time of writing my third edition no pre-Cretaceous ankylosaur had been identified. *Scelidosaurus* of the Lias is a small quadrupedal armored form, generally assigned to the Stegosauria. However, this form is incompletely known, the type of armor (whether stegosaurian or ankylosaurian in arrangement) is uncertain, and, most particularly, the pubis—the most diagnostic element in the skeleton—was unknown. I was therefore pleased to be shown in the British Museum in the autumn of 1966 a new specimen in which the pubis is present. The posterior ramus, unlike the condition in typical ankylosaurs, is still present in primitive fashion, but the anterior process is reduced to a nubbin. *Scelidosaurus* is hence a primitive ankylosaur.

The only notable advance in our knowledge of ceratopsians in recent years is the description by Sternberg (354) of *Pachyrhinosaurus*, which has, instead of a horn, a peculiar crater-like structure on its forehead. Recent general reviews of ceratopsian evolution have been published by Colbert (350) and Sternberg (1949).

13

Birds

Relatively little advance has been made recently in avian paleontology. The subject is a difficult one. Bird bones in general are delicate and poorly preserved, so that good specimens are rarely come by. Further, in most groups diagnostic characters are intelligible only to experts with a thorough knowledge of avian osteology; and ornithologists who penetrate beneath the feathers are rare. Wetmore (1956) has continued work on American Tertiary fossils; on the American Pacific Coast, Howard has continued active despite recent formal retirement; and the late Alden Miller made contributions in this area, as his father Loye Miller, had long done previously. An active newcomer to the field is Brodkorb, who has published extensively during the past two decades. No general survey of avian paleontology had been made since that of Lambrecht (408) a quarter of a century ago, and that compendium, apart from being none too critical and accurate in certain regards, is badly out of date. Highly welcome, therefore, is a new catalogue of fossil birds by Brodkorb (407). Brodkorb kindly gave me access to the manuscript of the portions of this work not yet published, and with few exceptions I have availed myself, in generic listings and classification, of his valuable work. Apart from a few studies noted later, there may be mentioned general surveys of fossil birds by Wetmore (410), by Swinton (409), and (with emphasis on Pacific Coast forms) by Howard (411); a survey of fossil Anseriformes by Howard (426); a volume on moas by Oliver (1949); a study of Pliocene South American phororhacoids by Patterson and Kraglievich (422); and the discovery of new "pseudo-toothed" birds by Howard (1957).

Archaeopteryx.—This Jurassic bird still stands in splendid isolation; we know no more of its presumed thecodont ancestry nor of its relation to later "proper" birds than before. De Beer (1954) has published a restudy of the British Museum specimen, and among other conclusions, points out (happily) that *Archaeopteryx* generically is one and indivisible, and that the claim of Petronievics (1950) that the Berlin specimen represents a separate genus *Archaeornis* is without substantiation. A third specimen has recently, after a gap of two-thirds of a century, come to light in the lithographic limestone deposits (Heller 1960), but this adds little to our knowledge.

Cretaceous birds.—Since the days of the publication of Marsh's impressive monograph (417) on the "Odontornithes" it has been almost universally assumed that in general Cretaceous birds were probably all still furnished with dental equipment. But despite the persistent paucity of materials other than those of *Hesperornis* and *Ichthyornis* from the Kansas chalk, the matter is currently open to doubt. *Hesperornis*, as proved by articulated material, is definitely toothed, and it is curious, as Gregory (418) points out, that its jaws show considerable convergence to the pattern found in the bird's prominent reptilian contemporaries, the mosasaurs. *Ichthyornis* is another matter. A reexamination of the evidence led to serious doubts (Gregory 1952). Marsh restored *Ichthyornis* with a toothed jaw, but there is no actual association of its supposed jaw with identifiable *Ichthyornis* material, and Marsh originally gave a separate generic name to the jaw material. Not improbably the assignment of the jaw to the bird was entirely arbitrary on Marsh's part (unfortunately Harger, who is said to have written most of the Marsh monograph, did not live to tell the tale). Like that of *Hesperornis*, the jaw assigned to *Ichthyornis* is similar to that of mosasaurs. But in this case, Gregory notes, the similarity is so great that he suspects this supposed bird jaw is actually that of a baby mosasaur (an attribution, however, that is still open to doubt). Adding to our doubts about the general "odontornithian" nature of the Cretaceous avifauna is the description by Brodkorb (1963) of a series of remains of late Cretaceous birds which seem clearly assignable to several of the "modern" orders.

Ratites.—The great series of ratites—ostriches, emus, cassowaries and rheas, and comparable extinct types, notably the moas of New Zealand and *Aepyornis* of Madagascar—have long been objects of interest and controversy. Despite earlier advocacy of the point of view that the ratites are primarily flightless forms, few today would maintain this point of view; quite certainly they are derived from flying ancestors. A second and more important question is whether the ratites are a natural group, or whether it may not be that they represent a number of types of diverse ancestry which have abandoned flight for terrestrial life in parallel fashion. Many characteristics of the ratites, such as reduced wings, fluffy feathers, powerful hind legs, and so on, are features which might be expected in any large ground dweller, no matter what its ancestry. One common feature, however, appeared to be universally present, a "palaeognathous" structure of the palate, not necessarily primitive, as the name suggests, but certainly different from that found in any other birds except the South American tinamous (fliers, but rather poor fliers) which, it has been thought, may represent broadly the ancestral stock. But McDowell, in a paper published in 1948 in which he pointed out various differences between the palatal structures in the different ratites, appeared to have done away with this one substantial reason for belief in ratite unity.

Further reflection tends to revive belief in ratite unity, however (cf. Hofer 1954; Simonetta 1961; Bock 1963; Meise 1963). McDowell stressed the *differences* between the various ratite palates. What about the features of *resemblance* between their palatal structures? Looking at the matter from this point of view, it is apparent that (despite fairly great divergence in the ostrich) the ratites do have a common palatal structure, peculiar to them and differing notably from types seen in other bird groups. We can, therefore, still maintain the Palaeognathae—ratites plus tinamous—as a probably monophyletic grouping (despite the fact that fossil evidence for derivation from any early type is conspicuously lacking).

Penguins.—These intriguing birds have always been of much popular as well as scientific interest, and we have long known some fossil forms from Southern Hemisphere sites. In recent decades Simpson (419, 420) and Marples (1952) have published substantial works on fossil penguins. Simpson is to be blamed for demolishing a romantic suggestion as to penguin history put forth by Matthew (435). The penguins, in popular belief, are primarily antarctic birds. Matthew suggested that the penguin ancestors had settled on Antarctica in the early Tertiary times when that continent was habitable, lost the power of flight, and then, with the coming of late Tertiary and Pleistocene cold, had been pushed off the continent by ice and taken to the sea. Simpson, however, points out that the penguins are primarily south temperate zone, rather than antarctic, in major distribution and that in all probability their origin (like that of the auks of the Northern Hemisphere) is by direct descent from oceanic flying birds, without intermediary antarctic residence.

14

Mammal-Like Reptiles

Synapsid classification.—Despite the claims to early divergence from the basal reptile stock of such forms as the Chelonia, it is certain that the initiation of a discrete synapsid stock was a very early event in reptilian history. In the early Permian, synapsids are highly developed and diversified and had obviously already gone through a long antecedent history, whereas in beds of this age members of any other reptile group (cotylosaurs, of course, apart) are still very rare; and Carroll (1966) has recently found pelycosaurian remains in the oldest beds (those of the Joggins) in which reptilian remains of any sort are known. This early separation of the synapsid line leading to mammals should be kept firmly in mind by comparative workers seeking to find antecedent homologues of "soft" structures of mammals among modern reptiles. If the mammal ancestors diverged from those of other reptiles at the very dawn of reptilian history, some 300 million or so years ago, it is very unlikely that structural features of modern reptiles can offer any safe clue as to conditions in the ultimate reptilian ancestors of the mammals.

Brink (1963) has proposed that because of their very distinctive nature the Synapsida be raised from subclass to class status. While I can sympathize with this suggestion regarding an important and distinctive group, in which Brink's interests are concentrated and on which he is a very able worker, I can see no great advantage to be gained by this increase in the number of tetrapod classes, any more than, for example, I can see (as already discussed) a necessity for the erection of a class Batrachosauria for a group having a comparable relation to reptile ancestry.

In the case of other major reptilian groups, their derivation from primitive "stem reptiles" is usually far from clear. In the case of the synapsids, however, derivation from the captorhinomorph cotylosaurs is obvious. Some of the later and better-known captorhinomorphs are obviously specialized in certain regards (notably dentition). But as far as we know there is little, save the excavation of the temporal opening, to distinguish pelycosaurs (particularly ophiacodonts) from the more generalized captorhinomorphs such as *Romeria* or *Proterothyris*.

Perhaps because their divergence from other reptiles occurred far back in the Carboniferous, the boundaries between the Synapsida and other reptiles, apart from the cotylosaurs, are sharply defined. Only in a few forms is there any question of whether a given fossil is or is not a member of this group. The Mesosauria, as noted elsewhere, have sometimes been included in this subclass because of the supposed presence in *Mesosaurus* of a lateral temporal opening. There is no other feature suggesting relationship, however, and it seems probable that, as noted before, we are dealing with an independent twig from the basic reptilian stock. The only other animal of note which occupies a problematical position is *Petrolacosaurus* from the late Carboniferous of Kansas (Peabody 1952). This is a small reptile of rather generalized and primitive structure. The temporal region is imperfectly preserved. Peabody believed that two temporal openings were present, and that hence we are dealing with a primitive diapsid. I have examined this material on several occasions and find myself unable to be convinced that this is the case. Without positive proof of the diapsid condition being present, it is difficult to place *Petrolacosaurus* in the diapsid pedigree. The "timing" is wrong; there are no further traces of diapsids until a full period later (the late Permian). Although it is possible that diapsids evolved in the Carboniferous and then went into hiding for many millions of years before reappearing, one would like more positive evidence. If one disregards the dubious evidence concerning the temporal region, one sees in *Petrolacosaurus* a rather generalized primitive reptile that in no regard violated the characters seen in pelycosaurs; further, certain details of the postcranial skeleton are distinctive of edaphosaurian pelycosaurs, and it seems probable that we have here a member of the primitive edaphosaurian family Nitosauridae, which must have been in existence at this time.

Pelycosaurs.—Although this boundary has begun to fade, it has been and still is practical to divide the synapsids into two major groups, nearly completely separated in space and time: the Pelycosauria and Therapsida. The pelycosaurs are essentially primitive in nature, are mainly characteristic of American and western European deposits, and in time are almost entirely confined to the late Carboniferous and early Permian; the Therapsida are advanced in structure, were long mainly known from South Africa, and in time appear only in the middle Permian. Distinctions of time and area are now beginning to break down to some extent, but we can still deal separately with these two great orders, representing two successive radiations of this major evolutionary stock.

In 1940 I published, with the able assistance of L. I. Price, a review of the pelycosaurs based on a thorough study of all material then known (379). A few new forms have since been described, but apart from *Petrolacosaurus*, mentioned above, they do not add much to the picture, and I think the general story of pelycosaur evolution and classification set forth at that time is still valid. There are apparently three well-defined subordinal groups, Ophiacodontia, Sphenacodontia, Edaphosauria, the first consisting essentially of primitive, persistently amphibious, fish-eaters, the second advanced carnivores, the third a herbivorous side branch. Each tended to produce specialized end forms, and in all three there is in many cases a trend toward increase in size over the relatively short period of geologic time during which they are adequately known. On the other hand, in all three there are small primitive forms whose relationships can be made out by technical diagnostic features, but which are quite similar in general features.

There is a strong trend in sphenacodonts and edaphosaurians toward lengthening of neural spines and development of a dorsal "sail." Apparently this development took place in parallel fashion in four or five lines. Its function seems to be clearly a primitive essay in temperature regulation (Romer 380); the sail grows disproportionately tall in large species of *Dimetrodon*, with the result that the ratio of its surface area to body volume remains constant. The "sail" is most familiar in *Dimetrodon* and *Edaphosaurus*; but in edaphosaurs its development may have occurred independently in *Lupeosaurus* (in which Lower

Permian Texas genus, despite repeated search, the head is still unknown), and in the sphenacodontids some trend toward elongation is present in *Sphenacodon* and a high development of flattened spines is seen in *Ctenospondylus.* Problematical points regarding pelycosaurs have to do with (1) the nature and relationships of the forms lumped by Romer and Price in the Eothyrididae, and (2) the relationships of the caseids to the edaphosaurids.

Some slight development of a "canine" region is present in ophiacodontid pelycosaurs as in many early reptiles and amphibians. In the predaceous sphenacodonts, the upper canines are highly developed; but in addition, there are known (from incomplete remains) a number of other pelycosaurs which have large canines, but are not sphenacodonts, and seem to represent independent developments from a primitive ophiacodontoid stock. These I brought together in 1940 (379) as the family Eothyrididae. Several of the forms included show definitely ophiacodontoid postcranial features. *Eothyris,* however, is known only from the skull and was combined with them simply on the basis of its greatly developed canines (together with lack of evidence of sphenacodont relationships). It is perhaps unfortunate that I selected *Eothyris* as the type genus of this family. Langston (1965*a*) has suggested that *Eothyris* (and this family in general) is related to caseid ancestry. This is highly improbable; caseids are notable for canine reduction, whereas the canines of *Eothyris* are the most highly developed of any pelycosaurs; further, the postcranial features seen in other eothyridids are those of ophiacodonts, and not at all suggestive of caseids.

The Caseidae were the last pelycosaurs to develop—and to flourish. They were long known only from *Casea,* a small form found in a locality now known to be at about the base of the Vale Formation of the Clear Fork group, and thus from a higher position stratigraphically than almost any other vertebrate known, until Olson's recent work in the Texas Permian. (The "classical" Clear Fork fauna is from the underlying Arroyo Formation.) In recent years our knowledge of the group has increased greatly. Possible primitive caseids or caseid ancestors include *Colobomycter* (Vaughn 1958), the possibly identical *Delorhynchus* (Fox 1962), and *Oedaleops* (Langston 1965*a*).

One of the startling events in pelycosaur studies was the finding by Stovall in 1940 of the giant caseid *Cotylorhynchus* in the Oklahoma Permian (at long last a skeletal description has just been published (Stovall, Price, and Romer 1966). The seeming lack of variety in caseids in earlier work on pelycosaurs was apparently due to the fact that the caseids blossomed late; Olson's work in discovering faunas in horizons above the early Clear Fork levels has resulted in finding in higher Texas beds not only further species of *Cotylorhynchus*, but several new genera—*Caseoides*, *Caseopsis*, *Angelosaurus* (Olson 383)—and the caseids are the one pelycosaurian type to have surely crossed the stratigraphic and topographic boundary between the early American Permian pelycosaur horizons and those of Russia containing therapsids, for *Ennatosaurus* of the Russian Kazanian is definitely a surviving caseid (Efremov 1956*a*).

Reig (1967), in an interesting discussion of thecodonts, suggests that the varanopsid pelycosaurs might represent ancestors of the Archosauria. This I very much doubt. Once thought to be a form related to the Ophiacodontia, I found in preparing the "Review of the Pelycosauria" (379) that in point after point the postcranial skeleton of *Varanops* checked exceedingly closely with the sphenacodonts and obviously represented a stage antecedent to them. The sphenacodonts are clearly ancestral to therapsids and, through them, to mammals. That archosaurs and mammals spring from a single pelycosaur stock seems highly improbable. Reig's argument springs primarily from the presence of an antorbital fenestra in the genus *Varanodon* (Olson 1965). Other seeming points of resemblance between varanopsids and a potential archosaur ancestor are in general merely features which varanopsids, as relatively primitive pelycosaurs, have retained from captorhinomorph ancestors. And that archosaurs are of captorhinomorph ancestry is, of course, highly probable.

Therapsids.—As with the pelycosaurs, this very extensive and varied group is clearly marked off from other non-synapsid reptilian groups, although at the base we are now beginning to find some forms close to the pelycosaur-therapsid boundary, and at the top we are increasingly finding problematical forms lying along the therapsid-mammal boundary. Evolved from pelycosaurs at about the beginning of middle Permian times, therapsids increased rapidly in numbers and variety, so that (for example) in the late Permian of South Africa they constitute over 90 per cent of all specimens found. They continued abundant in the early and (as we now know) middle Triassic, only to dwindle in significance toward the end of that period and vanish in the Jurassic.

For a long time the Karroo System of South Africa was the source of most of our material of therapsids. Despite his lack of conversion to evolution, Sir Richard Owen, by whom much of the early material was described, saw clearly that we were dealing with forms morphologically antecedent to mammals. Huxley and Seeley worked to some extent on South African therapsids, but it was Broom, that perceptive (if hasty and somewhat superficial) worker of genius, who, aided by studies in greater depth by Watson, brought the therapsids to worldwide attention. Later such men as Haughton, Boonstra, Huene, and more recently a series of younger men, notably Brink and Crompton, entered the field. An important event of recent years was the establishment of the Bernard Price Institute in Johannesburg which, with Brink in scientific charge and Kitching as a superb collector, has made the collection and description of Karroo therapsids a major objective. In recent decades, work based on Parrington's and Huene's collections from East Africa have somewhat expanded the African picture, and a recent English expedition (Attridge *et al.* 1964) promises further contributions to our knowledge of forms from Tanzania and Zambia.

Of other continental areas, little therapsid material has come from western Europe or from North America, apart from recent discoveries by Olson (383) in the American Permian which suggest the presence here of the initial stage of therapsid development; begun in Texas, Olson's work is continuing, with promise of good results, in Oklahoma. Russian workers have in recent times added greatly to our knowledge of the Permian therapsid story. More than a century ago a few tantalizing fragments were collected in the Russian copper beds; at the end of the last century Amalitsky collected considerable materials from northern Russia showing the presence there of a late Permian fauna comparable to that of South Africa. Much more important, however, have been discoveries in the region west of the Urals in the last three decades, where, in addition to some later material, there have been numerous finds of relatively primitive therapsids in horizons (Russian zones I-II) which appear to be, in great measure, older than the oldest of Karroo fossiliferous zones (the *Tapinocephalus* Zone of the lower Beaufort). Efremov and Orlov described much of this material, and studies are being continued by younger workers, notably Chudinov and Vjushkov. A large part of the results have been summarized by Olson (383).

South America, further, is currently coming rather strongly into the limelight. The Santa Maria beds of southern Brazil contain a fauna including therapsids of middle Triassic age, mainly described by Huene (238). Similar beds have more recently come to light in the Ischigualasto region of Argentina (Romer 1962*b*, and descriptive papers by Reig, Bonaparte, Cox, etc.); a somewhat earlier Triassic therapsid fauna has been found in the nearby Chañares region (Romer 1966, 1967*a*; Romer and Jensen 1966); and Bonaparte has very recently found earlier Triassic therapsids.

A very small amount of therapsid material had long been known from southern Asia. Recently early and late Triassic faunas, principally described by Young, have been discovered in China.

In earlier decades of the century, constant increases in our knowledge of therapsid types were made, principally from South Africa, and most notably by Broom. Broom's work ended in 1949, but the tempo of publication on therapsids has not diminished but actually increased. In 1945, after over a century of collecting, about 230 genera of therapsids had been discovered; in the last two decades about 110 additional forms have been described.

It has long been generally agreed that the therapsids are of pelycosaurian origin, and in the review by Romer and Price in 1940 (379) it was, I think, made clear that it is specifically the Sphenacodontidae, in the shape of short-spined relatives of *Dimetrodon*, that are the point of origin. Olson (383), in agreement with current tendencies to suggest polyphyletic origins for many groups, has suggested the possibility that dinocephalians and anomodonts have risen independently from the caseid pelycosaur stock. I feel unable to believe that weight should be given to this suggestion. There seems to be among dinocephalians almost every degree of transition leading from primitive types similar to and obviously rather closely related to sphenacodonts on to carnivorous and thence to herbivorous dinocephalians. It is the more advanced, not the primitive dinocephalians which show some analogy to caseids—forms with somewhat similar habitus. A key character upon which I believe great weight may be placed is the peculiar flange of the articular bone of the jaw, developed among pelycosaurs only in the sphenacodontids and retained in all primitive (and many advanced) therapsids—dinocephalians and dicynodonts, as well as theriodonts. It seems highly unlikely that this odd feature, so characteristic of sphenacodonts and obviously directly transmitted to at least certain primitive therapsids, should have developed in the descendants of caseids, in which the jaw structure is quite different from that of sphenacodonts.

Through the work of Broom and Watson, particularly, the nature of a number of assemblages of therapsids became clear several decades ago. Among carnivorous types there could be distinguished the Gorgonopsia, primitive in many ways but with a reduced dentition; the Therocephalia, in most regards equally primitive but with such diagnostic features as large palatal fenestrae; the Bauriamorpha, therocephalian descendants in the Triassic with a secondary palate and other advanced features; the Cynodontia, bauriamorph contemporaries of even more advanced structure; the Dinocephalia, middle Permian forms primitive in many ways but with a trend toward large size and herbivorous habits; the Dromasauria, represented by a few small and poorly known Permian specimens; and finally, the great group Dicynodontia, turtle-billed herbivores highly abundant in the Permian and persisting until the late Triassic.

It has long been agreed that the first four groups—all essentially carnivorous—could be united as a suborder Theriodontia, but beyond this there has been until recently little attempt to work out a more general classification or phylogeny of the therapsid group. A decade or so ago Watson and I (381) agreed jointly to work out a classification (and inferred phylogeny) with particular attention to the theriodonts; Watson, whose memory was remarkable and who had seen almost every specimen, was responsible for family assignment; the two of us together took responsibility for major elements in classification and relationships.[1]

A major novelty was our assumption that there occurred early in therapsid evolution a cleavage into a carnivorous and a herbivorous branch. For the former, the term Theriodontia was already available, and in it we included without question the gorgonopsians, cynodonts, therocephalians, and bauriamorphs. For the herbivorous branch of the trunk we utilized the term Anomodontia (sometimes used as a synonym of the Dicynodontia alone) and in this group included the herbivorous dinocephalians, dromasaurs, and dicynodonts. At first sight, these forms seem quite divergent from one another, but the Russian genus *Venyukovia* spans the gap between herbivorous dinocephalians and dicynodonts, and the dromasaurs apparently fit into the general pattern of the anomodont group, which it appears reasonable to believe is a natural unit. The herbivorous dinocephalians obviously belong in the Anomodontia; but since some of the Russian carnivorous dinocephalians show a very primitive structure, we decided to split the group, placing the herbivores as Dinocephalia proper among the Anomodontia; the carnivores, on the other hand, we assigned, as Titanosuchia, to the base of the Theriodontia (and, by implication, at the base of the Therapsida as a whole). Among the theriodonts, the four classic groups were retained, but since more "advanced" families formerly included among the Therocephalia (such as the scaloposaurids) seem to be leading toward the typical bauriamorphs, we included them in the Bauriamorpha, rather than the Therocephalia. A number of advanced (and dubious) forms from the late Triassic, including the peculiar tritylodonts, were placed by us in a theriodont infraorder Ictidosauria—a Broom term based on two then unnamed advanced late Triassic skulls in the Bloemfontein Museum. This classification was not, of course, aimed to be a definitive one, but a working base from which future departures might be made. For the most part it still appears reasonable and acceptable. But it is, of course, highly susceptible to modification and improvement.

[1] Sigogneau (1963) has complained, with justice, that in the gorgonopsians our family definitions were so restrictive that a large fraction of gorgonopsian genera could not be assigned to our families.

Brink (1960, 1963) accepted in part the Watson-Romer cleavage of the Therocephalia, agreeing that the ictidosuchids and scaloposaurids be united with the *Bauria* group infraordinally, but preferring to use the term Scaloposauria rather than Bauriamorpha for this assemblage. Rozhdestvenskii and Tatarinov in the amphibian-reptile volume of the new Russian Treatise on paleontology (1964) adopt a similar cleavage, but use suprafamilial terms—Scylacosauroidea and Ictidosuchoidea—rather than Therocephalia and Bauriamorpha. Since Crompton's description of one of Broom's "ictidosaurian" skulls as *Diarthrognathus*, discussed below, it is clear that the tritylodonts are not members of the same group, and in my current text, I have considered them a distinct infraorder Tritylodontoidea of the theriodonts, very probably of cynodont derivation; Rozhdestvenskii and Tatarinov have followed a similar course. These same authors have followed the Watson-Romer system in part in uniting dromasaurs and the *Venyukovia* group in the Anomodontia in a broad use of that term; they have not, however, followed us in placing the carnivorous dinocephalians in the theriodonts and the herbivores in the Anomodontia; instead, they adopt the conservative point of view of retaining the Dinocephalia as a unit-group, as one of the primary subdivisions of the Therapsida.

In this I believe they are in great measure correct, but to some extent wrong. Watson and I assumed, mainly on the basis of South African forms, that the Titanosuchia, including all carnivorous dinocephalians, must include the base of all the Therapsida. But if one reviews the cranial structure of titanosuchians, it becomes evident that nearly all forms often included in the group show a trend for the development of the "shouldered" chisel-like incisors characteristic of the herbivorous dinocephalians. They are hence presumably ancestral to these tapinocephalians and thus properly included in the Anomodontia in the sense in which Watson and I used this term. More primitive than any titanosuchian, seemingly close to the sphenacodont ancestor of the therapsids, is *Phthinosuchus* of Efremov (1954*b*). In our classification Watson and I included this genus in the Gorgonopsia (as do Rozhdestvenskii and Tatarinov), for the reason that the gorgonopsians are in most regards the most primitive of carnivorous therapsids. But *Phthinosuchus* lacks the reduced dentition and a few other features which debar the gorgonopsians proper from occupying a truly basal position among therapsids. I therefore later (382) proposed that this form be made the type of a group of "stem therapsids," Phthinosuchia, to which *Eotitanosuchus* and its close relatives (Chudinov 1960) may belong as well. Boonstra (387), before seeing my 1961 paper, reached a similar conclusion, separating out a primitive therapsid group for which he used the term Eotitanosuchia.

The ancestry of the Cynodontia is a problem of interest. Older and more primitive than the typical cynodonts of the *Cynognathus* Zone are late Permian forms such as the Procynosuchidae and Thrinaxodontidae (Galesauridae), the former particularly being very primitive in a number of features such as the incompletely known palate. One would tend to look for cynodont ancestors to the Gorgonopsia, the general skull pattern of which is a primitive type reasonably antecedent to that of the cynodonts; in all known gorgonopsians, however, the cheek teeth are too reduced to be ancestral to the dentition seen in cynodonts. At first sight, the therocephalians do not appear appropriate as ancestors. In such members of the group in which the feet are known, there is uniformly a reduction to the mammalian formula, whereas the cynodonts retain the "supernumerary" phalanges; further, typical therocephalians (and their bauriamorph descendants) are characterized by large palatal vacuities, unknown in cynodonts. But Brink (1951) has called attention to an obscure group of Permian forms such as *Silphedestes*, usually included in the Therocephalia in older classifications and certainly having affinities to the scaloposauroids, and proposed them as ancestors of the cynodonts. The known members of this group are specialized in some regards and rather incompletely known, and Watson and I in 1956 (381) were not disposed to accept Brink's conclusions.

Brink, in 1960, described a new form, *Scalopocynodon*, which he considered to belong to the silphedestid-scaloposaurid group. Yet it shows a palate definitely of the type present in primitive cynodonts, such as *Procynosuchus*, and thus strengthened his belief in the derivation of cynodonts from ancestral therocephalians. That *Scalopocynodon* is a primitive cynodont seems clear; but I may question whether it actually has anything to do with the silphedestid-scaloposaurid group of therocephalians. The seeming—but not certain—lack of a postorbital bar and zygomatic arch appears to have been the original reason for assuming its silphedestid-scaloposaurid relationships. Beyond this, features cited for such an association do not appear to be very strong in nature and certain ones are probably a retention of generalized primitive therapsid characters. On the other hand, the skull is most unscaloposauroid in most regards, such as the short tooth-row, strong cuspidate cheek teeth and, especially, lack of the palatal fenestrae characteristic of all therocephalians and bauriamorphs (except the Whaitsiidae). *Scalopocynodon* appears to be a good, if primitive, cynodont; but I do not think it contributes in any way to the theory of therocephalian ancestry for the cynodonts.

Subsequent to the publication of the Watson-Romer paper, the skull of an ictidosaurian was described by Crompton (405, 406) as *Diarthrognathus*, which proved (as Broom had suspected) to be very advanced in nature and is especially interesting in being transitional between reptilian and mammalian types of jaw articulation. Its therapsid affinities are none too clear. Its teeth suggest those of the problematical haramiyids (microcleptids) of the Rhaetic, recently redescribed by Peyer (1956), and I have ventured to provisionally associate them. Whether to class *Diarthrognathus* as a reptile or mammal is a question. I have retained it in the Synapsida, but Crompton (406) has recently advocated its transfer to the Mammalia (although it seems doubtful if it is related to any other members of that class).

Long obscure, the tritylodonts have in the past two decades become rather well known, due to Kühne's discoveries of *Oligokyphus* in British fissure deposits (403), the description of *Bienotherium* from China (Hopson 404; Young 1947*a*), further South African specimens, and the finding of tritylodont materials, as yet undescribed, in the American Southwest. Although far removed from the cynodonts in many ways, the suggestion of Crompton and Ellenberger (1957) that they are gomphodont descendants is reasonable.

There have long been known from the *Cynognathus* Zone of South Africa a few cynodonts, such as *Diademodon*, in which "gomphodont" dentition of broadened molar patterns developed. A remarkable development in recent years has been the discovery of numerous gomphodonts in the Santa Maria and Ischigualasto formations of South America, in the Manda beds of East Africa, and in the Molteno of the Karroo. I summarized earlier knowledge of them in 1961 (Romer 382). Recently a highly varied and abundant series of early middle Triassic age has been found in the Chañares fauna of Argentina (Romer 1966, 1967*a*); Bonaparte (1966) has found earlier Argentinian gomphodonts; and Baird informs me that a late survivor is present in the early Newark beds of Nova Scotia.

On the anomodont side of the therapsid picture, Boonstra (387, 1962, 1963) has been the main student of the dinocephalians, which, apart from a few primitive Russian genera, are known only from the Middle Permian *Tapinocephalus* Zone of South Africa. Boonstra has brought evidence supporting the point of view that the Dinocephalia instead of being divisible into but two groups, should be split into four—Anteosauria, Titanosuchia, Tapinocephalia and (usually appended with doubt to the Gorgonopsia) the Styracocephalia. He points out that the typical titanosuchians are to be regarded as herbivores, not (as previously thought) predators. The reasons for the speedy extinction of this seemingly flourishing group are far from clear. Possibly it was due to competition from their dicynodont relatives, which flourished enormously in the late Permian. These forms, in turn, became greatly reduced in variety and numbers in the early Triassic, only a few (but widespread) large types persisting to late in the period. Here, perhaps, the rise of the gomphodonts and rhynchosaurs characteristic of the middle Triassic may have been responsible for the decline of the group. Cox (400) has recently made a valuable study of the large Triassic dicynodonts, but there is much to be done in unraveling the dicynodont story; for example, about 130 species have been described as pertaining to the typical genus *Dicynodon*. The discovery in the Russian Middle Permian of *Venyukovia* did much to explain the phyletic origin of the dicynodonts (Watson 401); the recent discovery of *Otsheria* (Chudinov 1960) has strengthened this conclusion, reached from *Venyukovia*, that the pedigree of the dicynodonts traces back to dinocephalian-like forms. A study of the highly specialized jaw mechanics by Crompton and Hotton is in preparation.

15

The Structure of Mammals

In my third edition this chapter remains essentially unchanged. Except for the monotremes, mammals show an essentially uniform structural pattern and this pattern has been well known in major features since the days of Owen. This is true, particularly, in the skeleton, and except for details of this or that feature subsequently discovered in fossil forms, the description of the skeleton in Flower's classic work (433) is just as valid today as when it was written the better part of a century ago.

Partly due to this relative uniformity, I think, has been the tendency of many workers on fossil mammals to concentrate their attention on dentitions (and to a much lesser extent on foot structure). Teeth being the most resistant portion of the skeleton, we have little but the dentition in the case of many fossil forms; further, the fact that teeth alone (particularly molars) will often enable one to identify the genus and even the species of a fossil gives a "short cut" in systematic and faunal work of which mammalian paleontologists have not been slow to take advantage. So great has been this concentration on dentition that I often accuse my "mammalian" colleagues, not without some degree of justice, of conceiving of mammals as consisting solely of molar teeth and of considering that mammalian evolution consisted of parent molar teeth giving birth to filial molar teeth and so on down through the ages.

This concentration on dentition has had the unfortunate effect of restricting the point of view of workers in the mammalian field so that even when other skeletal material is available, it tends to be neglected. This is notably the case for cranial—particularly basicranial and otic—structures. Watson, many years ago, pointed out that in the braincase we have a region much less liable to adaptational change than teeth or feet and, hence, one in which basal diagnostic characters might be sought with profit. But in many otherwise excellent descriptions of mammals, the braincase is totally neglected. An excellent example is Scott's (469) series of monographs on the White River faunas in which the skull figures show the basicranial region beautifully shaded but with little indication of structure. Some workers on fossil mammals—Simpson, Patterson, Hough, McDowell and McKenna, for example—have utilized braincase structure; but most have avoided this difficult if important terrain.

In the dentition, the Osbornian system of nomenclature of placental molar cusps (Osborn 1907) has been almost universally accepted (except in such cases as the rodents, where tooth patterns make a special system of names necessary). Acceptance of this nomenclature does not imply acceptance of any theory of dental evolution below the placental level. Even today, after decades of discussion, the homology of the placental cusp pattern with those found in pantotheres and other Mesozoic mammal groups is uncertain. In a recent review, Simpson (497) refrained from naming or homologizing cusps in these forms and used instead an arbitrary series of letters and numbers in each Mesozoic group.

16

Primitive Mammals

In any broad consideration of mammalian classification and phylogeny, in the first quarter of our century, Flower and Lydekker's (432) and Weber's (430) classic works were necessarily heavily relied upon, with careful attention to the opinions of Matthew in regard to fossil forms. Currently, Simpson's 1945 classification (431) is the base upon which one leans heavily. As a modern, connected account of mammals, living and fossil, Thenius and Hofer's (428) recent book is excellent; their treatment largely follows the outline of Simpson.

Boundaries of the class Mammalia.—It has long been clear in the minds of those working with mammal-like reptiles and primitive mammals that with increasing knowledge it would be impossible to draw a sharp line between Mammalia and Reptilia; for the time, the shift in the jaw joint and loss of the "extra" lower jaw elements could be used as an arbitrary point of division. The time has now come when this position cannot be held; as shown by *Diarthrognathus* (Crompton, 405, 406), both types of jaw joint can exist simultaneously, and, on the other hand, it is now apparent that reptilian jaw elements may persist in Mesozoic forms which have always been considered as definitely mammalian. This breakdown of a sharp barrier between the classes has, not unnaturally, brought forth a spate of papers advocating a shift in the grouping of higher vertebrates. One rearrangement suggested as a possibility but not positively advocated by several authors (such as Patterson and Olson 1962; Simpson 497) is to restrict the class Mammalia to the Theria (which appear to be a monophyletic group) and push the Mesozoic mammals (except the pantotheres and perhaps symmetrodonts as possible therian relatives) down among the Therapsida. Most other suggestions have been for pushing the limits of the Mammalia downward, to include the Therapsida (Van Valen 1960) or the therapsids plus part of the Pelycosauria (Reed 1960). But if such expansion is started, it is hard to know where to stop. The boundary between the therapsids and their pelycosaurian ancestors has become faint, and one might well include the Synapsida as a whole among the mammals, although this would mean the inclusion of primitive and ancient Carboniferous reptiles hardly distinct from the basal cotylosaurian stock.

A different approach is advocated by Brink (395) and Crusafont-Pairo (1962). Arguing that striking physiological advance is enough to merit class status, Brink would make separate classes of forms which have advanced from the orthodox reptilian condition to include not only a class Aves, but also a class Pterosauria and a class Therapsida. Crusafont-Pairo would similarly erect a separate class, the Ambulatilia, for the therapsids.

I do not believe that any marked advantage would be gained by any one of the proposed changes. After all, if and when paleontological knowledge progresses much farther, *all* possible class boundaries will disappear, and quite surely a certain degree of polyphylety will appear to be present in any evolutionary transition from class to class. On the whole, it would seem to me wisest to maintain essentially our present concept of the class Mammalia, although admitting that it is to some degree an artificial grouping (Romer 382; Simpson 1959, 1960*b*).

Mesozoic mammals.—Our advances in knowledge of Mesozoic mammals seem to be on a "quantum" basis rather than on one of gradual steady progress. Following a long period of relative quiescence, Simpson's major works of 1928 and 1929 (495, 496) brought our knowledge of the then available materials of the older mammal orders up to a new high level. Then followed nearly two decades of stagnation. Beginning in the 1940's a new phase of research and discovery opened with Kühne's (1958) demonstrating the possibilities of Rhaeto-Liassic fissure investigation ably followed up by Kermack and Mussett (description of a *Morganucodon* of similar age from China by Rigney and Kermack is eagerly awaited as this is written). We may cite among a variety of new discoveries: a late Triassic triconodont skull from China (Patterson and Olson 1962); a new Triassic form from South Africa (Crompton 1964); Patterson's fragmentary but important Texas materials (502) from the Lower Cretaceous, and further material from the Texas beds being collected and described by Slaughter (1965); much new material from the classic Lance sites described by Clemens (452) and a fabulous number of late Cretaceous specimens from Bug Creek, Montana, under study by Sloan and others (Sloan and Van Valen 451). Studies of such early mammalian material have received additional stimulus from the parallel studies on advanced therapsids mentioned earlier—studies which have brought into view the possibility (or rather probability) that our older mammalian assemblages may be polyphyletic. Perhaps in a few years we will have reached a period of relative stability on a new level far advanced over that possible at the time of Simpson's 1928–29 reviews. We are currently, however, in a condition of flux and considerable uncertainty. I have attempted in my text to present, as far as possible, a consensus of current opinion, a good deal of it well summarized, for example, in a 1962 paper by Simpson (497). But it is probable that much of the thesis here presented will be outmoded in a few years.

I have continued to follow long-established custom in considering the existing mammals as divisible into two subclasses, Prototheria for the monotremes and Theria for all others, with the Theria subidivded into Metatheria for the marsupials and Eutheria for the placentals. But attempts to weld the extinct Mesozoic orders into the classification of modern mammals have always presented a very considerable problem and still do, despite recent advances. The major divisions of the class are based primarily on the nature of embryological development, and of that of the extinct orders we know nothing. Much diagnostic data could be obtained from the "soft anatomy" of the groups concerned; and of this, again, we are in almost complete ignorance. Were even the skeleton well known, we could gain many valuable clues as to relationships; but here again we are, unfortunately, without adequate information in most cases. Of the multituberculates, the skull is known in some late forms, and the new Bug Creek material promises considerable information regarding the postcranial skeleton. But of other groups, we are abysmally ignorant of postcranial structure and with a very few exceptions are equally ignorant of the skull. We are still mainly confined to the erection of hypotheses based on dentitions and incomplete lower jaws.

As of two decades ago, known lower types included, in addition to the multituberculates, the problematical "Rhaetic" teeth of the "microlestids" or "microcleptids," and a series of jaws and teeth ranged in three orders—Triconodonta, Symmetrodonta, and Pantotheria (the Trituberculata of older classifications). Of the true nature of the "microlestids" we know little more than before, despite meticulous restudy of these little teeth by Peyer (1956); the fact that it has been necessary to rename them the Haramiyidae (Simpson 1947) is hardly an advance in knowledge. There are numerous further unsolved problems. Should the "ictidosaurian" *Diarthrognathus* be retained among the therapsids, as I have done in my text, or be shifted over to the Mammalia, as Crompton (406) now advocates? Is *Morganucodon* (which may be synonymous with *Eozostrodon*) related to the docodonts (a group which Patterson [502] pointed out as worthy of a separate ordinal position)? What is the position of the Jurassic *Amphilestes* group which, as Patterson and Olson (1962) point out, is distinct from typical triconodonts? What are the relationships of the multituberculates? Are the symmetrodonts truly related to pantotheres and hence to the Theria? (Cf. a very recent paper by Crompton and Jenkins [1967].) Where does the poorly known Asiatic Mesozoic *Endotherium* (Shikama 1947; Chow, 1953) belong?

Monotremes.—Little can be said of the monotremes. They possess, to be sure, basic technical characters qualifying them for admission to the Mammalia, but they obviously are on a lower evolutionary stage than any other living mammals. It seems highly probable that a number of characteristic mammalian characters were developing in the therapsids, and hence it has long been generally believed that they represent an independent development from an advanced therapsid stock. But because of their high degree of skeletal specialization and loss or reduction of teeth, there is little basis for considering them in connection with the paleontological story unless—or until—ancestral types are discovered in the Australian early Tertiary.

Gregory's "palimpsest theory" (1947), that the monotremes represent a regression from a higher evolutionary level, cannot be accepted, particularly because of the necessary assumption that egg-laying has been resumed secondarily. Kermack and Mussett (1958) for a time suggested that monotremes were related to the Mesozoic docodonts, but later Kermack (499) abandoned this idea. The complex molar pattern, with a multiplicity of minor cusps, found in some gomphodont cynodonts (Romer 1967*a*), suggests, broadly, comparison with the similarly complex cheek teeth of *Ornithorhynchus*; but no transitional types are known.

Marsupials.—Little in the way of important new information has come to light concerning marsupial history in the past two decades. Some new data have been gained regarding Australian Pleistocene forms, and Clemens (1966) has ably reviewed and revised the late Cretaceous opossums (a revision which appeared too late for me to utilize it in my text); Simpson (459) and Paula Couto (1952*a*) have added to our knowledge of early South American forms. Stirton (509, 510, etc.; cf. Woods 1962; Gill 1965) has made a brave attempt to push our knowledge of Australian forms back into the Tertiary. He has succeeded to a modest degree; but the forms reported belong to groups already familiar and shed no light on the pattern of radiation that presumably occurred in the early Tertiary. An interesting study is that of Patterson (507) on little *Necrolestes* from the Miocene of Patagonia. This had been generally considered to be an insectivore, but an insectivore badly out of reasonable position in time and space. Various authors had suggested that *Necrolestes* might be a marsupial; Patterson has shown that this is actually the case. A further contribution to the history of South American marsupials is Patterson's description (1952) of the very distinctive, rodent-like, caenolestoid *Groeberia*; *Microtragulus*, as yet not fully described, appears to be another specialized form, the two indicating a wide radiation among caenolestoids.

Among Australian marsupials, as among placentals in other continents, the Pleistocene was notable for the development of large forms now extinct. Some of these marsupials, such as *Diprotodon*, are obviously comparable to existing types;

Thylacoleo, however, with reduced molars and the last premolars developed into large shearing teeth, has no living relatives and its habits have been a subject of debate. It was early thought to be herbivorous, but the matter is still in doubt (Gill 1954; Woods 1956).

In early days the marsupials were almost universally considered to comprise two suborders, the more primitive Polyprotodonta, with a goodly supply of incisors and more or less carnivorous habits and, derived from them, the Diprotodonta, with more or less reduction of incisors and with a vegetarian diet; the bandicoots (Peramelidae) occupy a somewhat intermediate position. With the discovery in South America of the living *Caenolestes* and fossil relatives which are neither good polyprotodonts nor diprotodonts, the picture became more complex. Simpson in 1930 (503; cf. Simpson 431) proposed that the classic suborders be abandoned and that, instead, the marsupials be arrayed in six superfamilies. Three of these—Didelphoidea, Borhyaenoidea, and Dasyuroidea—represent the old Polyprotodonta; Simpson's Phalangeroidea are the renamed Diprotodonta; the Perameloidea are the bandicoots; the Caenolestoidea include the aberrant South American *Caenolestes* and its fossil relatives. In an interesting recent review of marsupial evolution, Ride (511) retains Simpson's six superfamilies, although suggesting a possible subordinal classification in which the three American superfamilies would form one suborder, the three Australian ones a second.

I see no advantage in splitting the erstwhile polyprotodonts into three discrete groups. It seems clear that the didelphodonts, the "opossums," are a central stock from which the remaining marsupial types have been derived, and the borhyaenids of South America and the dasyurids of Australia represent two groups which retain a considerable basic similarity to their didelphid ancestors. The close similarity between members of the two groups does not mean, as Sinclair (506) and Wood (1924) believed, that they are quite closely related (or that a South American–Australian land connection existed). But these similarities certainly indicate that we are dealing with two parallel evolutionary lines arising out of the didelphid stock. Ride tends to emphasize the fact of geographical separation, but this does not, to me, warrant placing the two groups in categories distinct from their didelphid ancestors; phylogenetic, not geographical, proximity is the point to be kept in mind.

Even if, however, the polyprotodont and diprotodont groups be maintained, there is, of course, strong reason for the erection of a separate group for *Caenolestes* and its fossil allies of South America. Like the diprotodonts, the caenolestids have departed far from the ancestral polyprotodont pattern. There are interesting parallels between the caenolestoids and diprotodonts, but although Ride wistfully suggests the possibility of there having been a common ancestor, there is no fossil evidence to support such a view and, as Ride himself points out, dental evidence tends to negate it.

The Peramelidae are a minor nuisance in marsupial classification. I have in my text placed them in an independent suborder. But apart from syndactyly they are essentially good members of the polyprotodont assemblage. There is, however, some trend toward syndactyly among polyprotodonts (as *Marmosa*); it might be best to consider peramelid syndactyly as due to parallelism and include the peramelids in the Polyprotodonta (Tate 1948).

Subsequent to the writing of the above discussion, I received a more recent review by Ride (1964). In this he advocates raising (quite unnecessarily, I think) the Marsupialia to a superorder; he divides them into four groups—Marsupicarnivora, Paucituberculata, Peramelina, and Diprotodonta. Except that my suborders are raised to orders and the names used are different, these four correspond precisely to the four groups I have used—Polyprotodonta, Caenolestoidia, Peramelida, and Diprotodonta. Ride and I appear to be in accord about marsupial evolution and systematic arrangement.

17

Placental Origins; Insectivores; Bats

Placental superordinal grouping.—Among reptiles, apart from a few short early twigs, most types can be readily grouped into a few large and seemingly natural superordinal or subclass assemblages, such as the Archosauria, Synapsida, and Lepidosauria. It would seem possible a priori that the great Tertiary radiation of a score or so of placental orders could be subjected to similar treatment, and two well-considered attempts at this have been made, by Gregory in 1910 and by Simpson in 1945 (431). Neither attempt, I feel, has been successful.

Gregory arranged the placentals in seven superorders as follows: Therictoidea, including Insectivora and Ferae (Carnivora); Archonta, with Menotyphla, Dermoptera, Chiroptera, and Primates; Rodentia, for the "true" rodents, as Simplicidentata, and the Duplicidentata (Lagomorpha); a questionable assemblage of Edentata, for the Taeniodonta, Tubulidentata, and Pholidota as well as the edentates proper as Xenarthra; Paraxonia for the Artiodactyla; Ungulata for all other ungulate orders; and the Cetacea as a superordinal group of their own.

Let us review the contents of these proposed superorders. By Gregory the carnivores are bracketed, as the Therictoidea with the Insectivora, apparently from the point of view that since the insectivores, the basal placental group, were basically eaters of animal food, the carnivores, as flesh-eaters, are the most direct offspring of the ancestral stock. But although the dentitions of some of the earliest carnivores are not too far removed from those of ancestral insectivores, transitional types are few; there appear to be few positive reasons for bracketing the insectivores with the carnivores rather than for associating them with any of the other placental orders, to all of which they presumably gave rise. Gregory's Archonta include primates and bats as its two main components. The "Menotyphla" are reasonably associated with the primates, since the tree-shrews, most notable of menotyphlans, are reasonably considered as potential ancestors of the Primates and are by many included in that order; the "flying lemurs" of the Dermoptera are presumably placed here because their "flying" ability suggests association with the Chiroptera. But there are few resemblances between Primates and Chiroptera (apart from the fact that both are insectivore descendants), and the Archonta seems an artificial group.

Gregory's group Rodentia, including the true rodents and the lagomorphs—the rabbits and hares—is equally artificial. To be sure, both groups were at one time lumped in a common order, but even at the time Gregory wrote it was becoming evident that the two types were very distinct; apart from the presence of chisel-like incisors (found in a variety of other mammals as well), there is practically no feature common to the true Rodentia (the former Simplicidentata) and the Lagomorpha (*olim* Duplicidentata), and it is highly probable that the two have quite separate pedigrees.

Gregory's Edentata includes, as well as the "proper" edentates (the South American Xenarthra and kin), the aardvarks, pangolins, and the short-lived taeniodonts of the early Tertiary. There are no common features except the tendencies for tooth specialization, and no hint of common pedigree. The oldest xenarthrans show no indications of taeniodont relationships; the ancestry of the pangolins is unknown; the aardvarks are quite possibly related to the condylarths.

Gregory separates the even-toed ungulates, as Paraxonia, from all other ungulates. This seems to have been due to the specializations of the artiodactyls and some suggestions that this group had affiliations with carnivores. Later research, however, has tended to show that the supposed carnivore affiliations are groundless, and that there is no more reason to exclude the artiodactyls from the general ungulate assemblage than to exclude any other hoofed mammal type.

We then come to the remaining "hoofed" orders, which Gregory includes in a superorder Ungulata. It may be reasonably asked whether all these forms are a truly natural, phylogenetic unit. The order Condylarthra, it is generally agreed, includes a variety of primitive early Tertiary forms which are rather surely ancestral to the Perissodactyla, the odd-toed ungulates, and to the extinct Litopterna of South America. But there is relatively little evidence of condylarth descent for many other important groups, such as the various archaic types of the early Tertiary, the subungulates, the great notoungulate group of South America, or the Artiodactyla. The characters common to all ungulates are few, apart from the fact that they are generally terrestrial placentals which have tended toward a purely herbivorous mode of life, have developed molar teeth capable of dealing with vegetable food, and in general have shifted from claws to hooves, with, frequently, a trend toward large size and adaptations for a running gait. It is possible that all of these forms have arisen from a single ancestral stock. But, in default of adequate proof to the contrary, it is equally likely that several types of early placentals moved toward a diet of plant food with consequent trends, in parallel fashion, toward the taking on of obvious "ungulate" specializations. It is far from certain that the Ungulata form a single phylogenetic unit.

The whales and porpoises, the Cetacea, are so distinct that Gregory is justified in assigning them a separate place in his system. But in doing this, there has been, of course, no erection of a new major group; an order has simply been raised to superordinal rank.

As has been seen above, and as Simpson has pointed out, Gregory's groupings are certainly unnatural in some regards and broadly open to suspicion in others. In 1945, Simpson (431), in parallel fashion, attempted to assemble the placental orders into a series of "cohorts," but despite my respect for his broad knowledge of early placentals, I find his groups as unsatisfactory as those of Gregory. Simpson arrayed the placentals in four cohorts: (1) Unguiculata, including Insectivora, Dermoptera, Chiroptera, Primates, Tillodontia, Taeniodontia, Edentata, and Pholidota; (2) Glires, for Rodentia and Lagomorpha; (3) Mutica, for the Cetacea; (4) Ferungulata, for the carnivores and all the ungulate and subungulate orders.

In the two cohorts of least complexity, Simpson exactly parallels Gregory. He raises the cetaceans to a superordinal rank as the cohort Mutica and unites, as the cohort Glires, the true Rodentia and the Lagomorpha—orders which, as mentioned above, show no real relationship to one another.

Simpson's suggestion that carnivores and all types of ungulates are members of a single major group, the cohort Ferungulata, is a stimulating and thought-provoking one. But this suggestion rests on little positive evidence, and there are disturbing factors. It appears to be based primarily (Simpson 431, p. 216) on the presence, in the earliest Tertiary, of forms seemingly structurally intermediate between carnivores and primitive ungulates. The Arctocyonidae and Mesonychidae, customarily included as "creodonts" in the order Carnivora, greatly resemble members of the primitive ungulate order Condylarthra. Recent work, however, indicates that these two families are best considered not merely similar to condylarths, but actually members of that order, not of the Carnivora, and not necessarily related in any fashion to proper carnivores except insofar as they are all descended ultimately from basal placental ancestors. I know of no particular resemblance of ungulates to the miacids, ancestral to the true Carnivora of the later Tertiary, nor to the typical creodonts of the hyaenodont and oxyaenid families which, as discussed later, may have had an independent pedigree.

All in all, there is thus currently little positive reason to believe that carnivores (*sensu lato*) plus ungulates form a natural unit. There is, furthermore, the underlying assumption that the ungulates are a monophyletic assemblage—an assumption which, as mentioned above, is open to considerable doubt. Much as I personally favor tidiness and simplicity in classification, I must regretfully doubt the reality of a "natural" ferungulate cohort.

The cohort Unguiculata has still less coherency. It is, essentially, a voluminous catchall in which have been placed everything that is not a whale, rodent, carnivore, or ungulate. If we take the Insectivora as a base, inclusion of the Chiroptera as an aerial offshoot from the insectivores is reasonable. Association of Primates and Insectivora seems to have been made chiefly on the grounds that clear differentiation of the Primates from the basal placental insectivore stock appears to have taken place slowly and confusingly, so that it is difficult in the fossil record, where dentitions are often all that is known, to distinguish between members of the two orders. Not only in earlier days, but currently as well, there are about as many cases in which there is doubt about whether a given form is an insectivore or primate, on the one hand, or a "carnivore" or ungulate, on the other. In the Paleocene most ordinal boundaries are blurred, and there seems to be no stronger reason to associate the insectivores more closely with their primate descendants than with their early carnivorous or proto-ungulate offspring.

Still less evidence is present for closely associating the edentates with Primates and Insectivora, Simpson merely stating (431, p. 190) the reasonable thesis that "the edentates arose from proto-Insectivora." True, but so arose, presumably, all other placental orders. Although the early palaeanodont edentates are primitive in many regards, such edentates as glyptodonts and ground sloths appear to have departed as widely from a primitive pattern as the whales or rodents which Simpson thinks worthy of segregating into distinct cohorts.

The Dermoptera, Taeniodontia, Tillodontia, and Pholidota are included in the Unguiculata, it would seem, mainly because they do not belong anywhere else. They are not carnivores, ungulates, whales, or rodents.

All in all, while suggested placental assemblages such as those of Gregory's superorders or Simpson's cohorts, are valuable in stimulating thought and research, it does not appear that at the present stage of knowledge we can successfully disentangle and reassemble the confused series of sprouts and shoots which constitute the early phyletic development of placentals. In contrast to the seemingly rather neat and tree-shaped pattern of phylogeny among the reptiles, that of placental evolution seems to be in the shape of a complex bush. I sometimes wish we were living in the Paleocene; we would then be able to place nearly all the placentals in a single order, and our difficulties of classification would be happily restricted to problems of families and genera.

Insectivora.—No order of mammals is as difficult to understand as the Insectivora. In most placental orders we have well-defined end forms which characterize the group, and from such forms we trace back through types in which *positive* ordinal characters are present, even if less and less distinct. Exactly the converse is true of the Insectivora; to consider whether a form should be included in this basal stock we look for the *absence* of outstanding characters which would define more "advanced" derivative orders; our approach must be a negative one. In considering the classification and limits of the Insectivora, I have profited by a number of discussions with McKenna and Van Valen, whose interests center about insectivores and other early and primitive placentals. But I hasten to say that they should not be held responsible for the results of my attempt to erect, for textbook purposes, a seemingly clear-cut (and surely deceptive) classification of the group. They are infinitely more competent than I am in this field, but would be the first to disclaim finality in this exceedingly complex area. McKenna's conclusions have been expressed in certain of his publications (464, 1960) and in a mimeographed (but not published) tentative classification of 1958. Van Valen, working first as a student and then as a postgraduate fellow in McKenna's laboratory, has roughly paralleled McKenna in his conclusions but has tended to express them more positively—and perhaps rather too brashly—in an attempt (mimeographed in 1965) at a new classification of insectivores.

What should be included in this order? To my mind, three principal components: (1) The typical living insectivores, such as the shrews, hedgehogs, and moles, and fossil genera definitely related to them. These are, of course, the forms upon which the order was primarily based by early workers in mammalogy, in days when fossil evidence was absent or of small account. (2) Truly primitive placentals, so generalized in nature that they can be considered as a basic placental stock and not members of any specific advanced order. (3) Forms which have departed from the placental base to some degree but which have not been so strongly modified in structure that they can be deemed worthy of having separate orders erected for them.

The typical living insectivores are, on the whole, a coherent and well-knit group to which the subordinal term Lipotyphla is reasonably applicable (begging, for the moment, the question of the zalambdodonts); McKenna, at least at one time (1960), would limit the term Insectivora to the "modern" forms. Most members can be placed in the modern families Erinaceidae, Talpidae, and Soricidae, although the Dimylidae have long been accorded family status as a mid-Tertiary side branch of the erinaceids. I have followed Van Valen in splitting off a series of primitive erinaceids as a separate family Adapisoricidae, and a few aberrant soricoids may deserve placement in a discrete family Plesiosoricidae.

These are the major stocks of modern insectivores, with a respectable Tertiary history, which in the case of the erinaceoids reaches back to the Paleocene and perhaps to the late Cretaceous. But these "dilambdodont lipotyphlans" are not to be considered a basal placental stock; all are more or less specialized and cannot be regarded in themselves ancestors of other placental groups. To find such potential ancestral types, we must look further down the placental tree. A group which has gradually come to be recognized as possibly occupying a position close to the placental stem is that of the Leptictidae; this family, although surviving until the Oligocene in the form of *Leptictis* and *Ictops*, traces back in the recently discovered *Procerberus* and the probably related *Zalambdalestes* to the Cretaceous. Early erinaceoids are not far removed structurally from the leptictids, and the two have often been included in the same superfamily (as, for example, in Simpson 431). But as Butler (514) pointed out, the leptictids are obviously a very primitive group which might be close to the basal placental stock. They lack, as far as is known, the "modernized" specializations of most lipotyphlans and, for example, show a molar pattern of a basal eutherian type, the upper molars being generally simply tritubercular with little trend toward hypocone development and with paracone and metacone well separated and well toward the outer margin of the tooth. It is not improbable that the leptictids may have given rise not only to later lipotyphlan insectivores and other short placental stems, but to condylarths and other ungulates, to primates, to creodonts—and, indeed, to later and more "advanced" placentals in general. Still further, it may be that, as discussed later, the deltatheridiid type of molar which gave rise to that of typical creodonts may be derived from the type seen here rather than from a Mesozoic-pantothere type of molar.

If early placentals are surveyed, a number of other forms appear also to lack observable specialization, including, for example, *Anagale* (which, as McKenna pointed out in 1963, is not a tupaiid, although fairly closely related to that family) and a number of rather obscure Paleocene and Eocene types noted for little except lack of specialization and absence of departure (as far as known) from the general leptictid pattern.

What should a group of apparently generalized and possibly ancestral placentals of this sort be termed? McKenna has at times (for example, 464, 1960) called them Menotyphla "faute de mieux," implying that at least they are not "modernized" lipotyphlans and considering them as a basal eutherian suborder distinct from the "typical" living insectivores. In my text I have probably overboldly brigaded them as a suborder Proteutheria, emphasizing their probable position at or close to the base of the Eutheria.

Not far removed from the presumed eutherian base may be the Pantolestidae, with distinctive as well as primitive characters. Somewhat farther removed, but perhaps also worthy of inclusion may be the Apatemyidae, usually placed among the primates (as in Simpson 431) but considered by McKenna (515) to be a somewhat rodent-like offshoot from the early insectivores. I follow Van Valen (1965*d*) in considering as a leptictoid offshoot the little extinct Eurasian group of Paroxyclaenidae.

And, finally, the Tupaiidae. The oriental tree shrew, *Tupaia*, and its relatives have long attracted much attention from students of mammalian classification and phylogeny. The tree shrews are certainly very primitive placentals. As first pointed out by Gregory (1910) and Carlsson (1922) they show a number of features which suggest they are ancestral to the primates. LeGros Clark, ever since his studies of 1924–26 on *Tupaia*, has in various publications (as 1934, etc.) emphasized the primate resemblances seen in tree shrews and advocated their inclusion among the primates. This has been rather generally accepted, and they were ranked in 1945 as members of the order Primates by Simpson (431) and (with unspoken reluctance) by me in the second edition of my text.

A long series of authors has, in recent decades, studied various points in tree-shrew anatomy in comparison with the primates. Their conclusions may be summed up in three categories: (1) In numerous regards tree shrews show primitive placental conditions from which structural features present in primates may have developed. (2) In certain areas tree shrews and primates show common positive characters not present in placentals generally. (3) In other regards tree shrews and primates differ clearly from one another.

The case for inclusion among the primates does not seem to be at all certain (cf. Van Valen 1965*c*). The first category merely emphasizes the primitiveness of tree shrews and indicates that they may represent ancestors of primates but may be ancestral to other placentals as well. The conflict of testimony in categories (2) and (3) mirrors the fact that, after all, modern tree shrews have had many tens of millions of years to evolve since eutherian history began and although remaining primitive in many regards, they may well have, during that time, made modifications of one sort or another either converging toward or diverging from primate descendants.

The tupaiids may be primate ancestors. But not improbably their importance is far greater. It may well be that in the tree shrews we see the most primitive of living placentals—forms not too distant from the common base of all eutherian stocks.

So far we have considered "modernized" insectivores and forms which may be considered as a basal eutherian stock. There remain for consideration a variety of other forms which are often placed in the Insectivora, including forms which are divergent from the common eutherian base but have not departed far enough to merit, I believe, erection of separate orders for their inclusion.

The elephant shrews of Africa, *Macroscelides* and its allies, have had a varied taxonomic history. They were long held by many workers, following Haeckel, to be related to the tree shrews, the two groups together constituting the suborder Menotyphla. In recent decades Evans (1942) alone has maintained this point of view, whereas other workers, beginning with Carlsson (1909), have tended to regard this grouping as artificial. The latter point of view seems currently the most reasonable. Little was known of elephant-shrew history until Patterson (518) pointed out that several African Tertiary types, formerly distributed among the marsupials, hyracoids, and mixodectid insectivores, were actually elephant shrews; the macroscelids are thus an ancient and diversified African group. Both Butler (514) and Patterson regard the macroscelid group as forming a separate order, but, in line with the "lumping" policy I fávor, I have included them in the Insectivora as the suborder Macroscelidea.

The "flying lemur," *Cynocephalus* or *Galeopithecus*, is customarily placed in a separate order, Dermoptera. There is actually little reason for this. The teeth are somewhat specialized, and there is a planing membrane. But the presence of such a membrane is hardly a feature of ordinal nature; after all, the development of a rather good structure of this sort is considered as worthy only of a subfamily cleavage among the squirrels. As Matthew and Granger (463) first pointed out in 1918, *Plagiomene* of the Eocene is surely an ancestral member of the group. The teeth of this form clearly show the derivation of the dermopteran dentition from a general basic "insectivore" pattern. There appears to be no more reason for splitting off the "flying lemur" into a separate order than for splitting off a number of other derivatives which are generally retained within the Insectivora.

Possibly related to the Dermoptera are the Paleocene Mixodectidae. Dubious as to position and often confused with the latter, are the Microsyopsidae.

Picrodus and *Zanycteris* are not bats, as is sometimes suggested, and are probably to be placed in the Insectivora "*incertae sedis*" (McGrew and Patterson 1962). Van Valen (1965*a*), on the basis of unpublished work, considers them to be primates, possibly tarsioids.

Problematical are the forms often bracketed as zalambdodonts, including *Tenrec* (*Centetes*) and other Madagascan forms; *Potamogale* of West Africa; the golden moles, such as *Chrysochloris*, of South Africa; *Solenodon* of the West Indies; and as possible relatives, *Apternodus* and a few other forms from the late Eocene and Oligocene of North America. The most prominent feature is the structure of the upper molar—narrow fore and aft, wedge-shaped, with para- and metacones far in from the outer border and partially or completely fused. Are these forms a unit group? Is this dentition primitive, proceeding directly from a Mesozoic pattern antedating the development of the typical placental trigon? If not, can they be derived from the *Deltatheridium* group of the Cretaceous and early Tertiary? Or, again, is the zalambdodont molar quite secondary, derived by modification in the Tertiary from ancestors with a more "normal" type of placental molar pattern? Possibly the zalambdodonts are a monophyletic group, but this is far from certain; Broom, decades ago, pointed out that the golden moles are quite distinct from the other African zalambdodonts in all regards except dentition. Few today would, I think, claim that the zalambdodonts are directly descended from Mesozoic pantotheres. Whether, however, they are directly derived dentally from the deltatheridian type, already present in the Cretaceous, is a point of view that can be debated. As noted elsewhere, Van Valen believes the deltatheridians to be ancestral to the typical creodonts. In his classification (543) of this group, which he makes a separate order of "Deltatheridia" (= Creodonta), he does not include the zalambdodonts but does in his text raise the question as to whether they "may or may not be included."

The third possibility remains—that they are related neither to Mesozoic types nor to deltatheridians, but have achieved their present molar pattern by modification from the typical placental trigon. We have little positive evidence regarding the ancestry of the living genera; the West Indian forms are known only from the Pleistocene, and while golden moles and tenrecids are present in the African Miocene (Butler and Hopwood 1957), modern characteristics were already established by that time. In the late Eocene and Oligocene of North America are a few forms, such as *Apternodus*, with "zalambdodont" dentition. It has not unreasonably been suggested that these genera are related to the West Indian zalambdodonts at least. McDowell (1958) would deny this. He, as well as Patterson (1957, 1962), however, believes that the zalambdodonts are very probably descended by Tertiary modification from lipotyphlan insectivores, presumably of the soricoid subgroup, rather than having had anything to do with deltatheridians or Mesozoic ancestral types.

Tillodonts and taeniodonts.—These two curious orphan groups are elements of an archaic early Tertiary placental radiation—elements which did not survive beyond the Eocene. They cannot, with certainty, be associated with any of the major, more successful, placental evolutionary lines but in their later members they have departed so far from the basal insectivore pattern that they are generally placed in discrete little orders. Gazin (521) has thoroughly reviewed the Tillodontia. He points out features which suggest relationship to early ungulates such as *Pantolambda* and *Arctocyon*. Van Valen (522) emphasizes the condylarth resemblances and advocates inclusion of the tillodonts in the order Condylarthra. Much as I favor tidiness in systematics, I do not feel that the evidence for this move is sufficiently strong.

Equally isolated are the Taeniodontia which, except for a recently discovered Asian Eocene survivor (Dehm and Oettingen-Spielberg 467), are exclusively North American and mainly Paleocene. The evolutionary story has been well reviewed by Patterson (520). There is no clue to relationship with other orders, and it would seem that they have arisen independently from an insectivore base.

Chiroptera.—The only addition of note in recent years to our knowledge of the Chiroptera is the description by Jepsen (1966) of an almost perfect skeleton from the early Eocene. This form is, however, a well-developed microchiropteran, differing in only a few points from "modern" types and giving us no insight into the history of the order. It is quite possible that certain forms assigned currently to the Insectivora are prechiropterans, but in the usual absence of postcranial materials, which might show whether or not the diagnostic wing structures were developing, such claims are hard to prove or disprove. Edinger (1964) has pointed out that the endocast of a Paleocene specimen in the Princeton collection shows a midbrain of chiropteran type, but Jepsen states that the skull concerned is that of a miacid carnivore; it is possible that sensory structures of chiropteran type, with associated brain modifications, may have developed more than once. Friant (1952) at one time claimed that the fruit bats were derived from microcleptids, but later retracted this improbable idea, and we have almost no paleontological evidence as to the origin of the Megachiroptera.

18

Primates

A very considerable amount of valuable work has been done on primate evolution in the last two decades. We may note, for example, various able reviews—particularly on the earlier and more primitive types—by Simpson (*526*, *1960a*), Simons (*525*, 1960), and McKenna (1965*b*), studies on North American Eocene forms by Gazin (*532*), and on various Eocene genera by Simons (*530*, *531*). Simons is collecting valuable material from the Fayûm which promises to shed new light on the early history of the higher Old World primates; Leakey's work in East Africa is making important additions to our knowledge of late Tertiary and Pleistocene forms from that region; LeGros Clark has continued to make major contributions to primate morphology and phylogeny (*523*, *524*, etc.). My volume went to press too early to include *Purgatorius* (Van Valen and Sloan 1965), now the oldest of known primates.

Lower groups.—As is well known, it is difficult to draw a boundary between the Insectivora and Primates among early Tertiary types. In many cases we have little data on the forms concerned except the dentition, and the fact that the primate dentition tends for the most part to remain persistently primitive and hence close to that expected among primitive insectivores adds to the difficulties. We have already discussed the tupaioid tree shrews and are here considering them as insectivores.

Particularly puzzling have been a number of forms, mainly Paleocene, which are to some degree primate-like but have specialized in the development of rodent-like incisors (cf. McKenna 1965*b*). *Plesiadapis* (most recently redescribed by Russell 461) is the best-known form; *Carpolestes*, *Phenacolemur*, and *Apatemys* are representatives of groups with comparable adaptations. I have followed current opinion (cf. Simons 525) in placing the families based on the three forms first mentioned in the Primates; the Apatemyidae, as noted earlier, are probably an analogous development within the Insectivora.

How to formally classify these three families is problematical. They presumably represent one or several side branches from the primitive primate stock. They do not appear to be related to the tarsioids; nor is there any reason to bracket them with the lemurs. I have provisionally classed them as a separate primate group, as a suborder Plesiadapoidea, although it may well prove eventually that we are dealing with several independently specialized types rather than a true phyletic unit.

Proper classification of the Primates is a difficult problem, to which, as Simpson (431, pp. 180–82) points out, there is no single obvious solution. The plesiadapoids (and tupaiids) apart, we are dealing with at least four distinct groups: (1) the lemurs, including early Tertiary forms and modern types, of which the latter can be subdivided into the less progressive Madagascan forms and the more advanced continental genera (galagos, lorises); (2) *Tarsius* and his early Tertiary relatives; (3) the platyrrhine South American monkeys; and (4) the catarrhine monkeys, apes, and man of the Old World. In the past it has been almost universal practice to unite the platyrrhines and catarrhines in a suborder Anthropoidea. With regard to the "lower" groups there has been variability. Lemuroids and tarsioids have been frequently separated into two subordinal groups; on the other hand, the two may be fused into a common "suborder Prosimii."

Simpson, in 1945, adopted the latter course and has been very generally followed. I do not believe that this procedure is a proper one; to lump all forms below the monkey-ape level as Prosimii is comparable to dividing all animals into Vertebrata and Invertebrata. Simpson justifies it primarily on the fact that early lemurs and the tarsioids are in some cases difficult to sort out. True; but this is to be expected—and even hoped for—if the order is a natural one; early members of two such groups should be convergent, no matter how far later members of the two have diverged.

And they have diverged far. Although we find that, in their isolation, the Madagascan lemurs developed odd and interestingly specialized types, the Lemuroidea are essentially a conservatively primitive group; the Tarsioidea, however, apart from giving rise early to ancestors of the living *Tarsius* with its locomotor specializations, evolved a progressive series of forms ascending toward the monkey-ape stage (an interesting advanced form of this sort is *Rooneyia*, recently described by Wilson [1966]). I do not believe that it is necessary to go as far as some workers have done (most recently, Hill 1955) in uniting the tarsioids with the more advanced primates as the Haplorhini, but I think it more in accord with their phylogenetic importance to separate them quite sharply, as a distinct suborder, from the very divergent lemuroids.

Of all the varied members of the Madagascan lemur assemblage, the most aberrant is the aye-aye, *Daubentonia* [*Chiromys*], in which, among other varied specializations, there has developed a set of rodent-like incisors rather comparable to those of the plesiadapoid group. Various writers have been tempted to consider the possibility that the aye-aye is actually related to the plesiadapoids; current majority opinion (which I have followed) is, however, that this is merely parallelism, and that *Daubentonia* has evolved from the lemurs proper.

I have, perhaps, been unduly conservative in placing all the Madagascan forms (apart from the aye-aye) in a single family Lemuridae, rather than splitting off the indris and avahis and their extinct relatives as the family Indridae, but the differences are not great. It is generally agreed that the continental lorisids—galagos, pottos, lorises—are a more highly evolved group than the Madagascan lemurs, and they are frequently separated from the Madagascan forms as a separate superfamily or infraorder. In most regards, however, they conform to the general lemur pattern; descent, paralleling the Madagascan forms, from adapid lemurs, although unproved, is reasonable. Until recently almost no fossil material of the lorisid was known. In a paper now in press Simpson (1967) shows that a variety of lorisids were already present in the African Miocene; group characters were then already fully developed, but these varied forms shed no light on group derivation.

I have followed Gazin (532) in dividing the Eocene tarsioids generally into Anaptomorphidae and Omomyidae and Simons (531) in placing several European types definitely in the Tarsiidae; I have not followed Gazin in dividing the Eocene lemurs proper into the parallel Old World adapid and New World notharctid families. I have (probably incorrectly) appended the Microsyopsidae, with doubt, to the tarsioids (cf. McKenna 1965*b*). Gregory was the first to point out that the tarsioids were not descended from the *Adapis-Notharctus* group but represent a more progressive parallel line.

Platyrrhines.—As said above, it has been nearly universal custom to include all higher primates—monkeys, apes, and men—in a single suborder Anthropoidea. Regretfully, it appears that the time has come when this term must be abandoned (Patterson 1954). One group of monkeys, the Platyrrhini, is confined to South America; the Catarrhini, including Old World monkeys, apes, and men, are known only in Eurasia and Africa. Unless (highly improbably) rafting of monkeys took place across the South Atlantic, any common ancestor of the two groups must have inhabited North America. But there is not the slightest trace of a higher primate in this continent, and the current consensus of opinion is that the two groups evolved in parallel fashion from tarsioids—some of which, such as *Rooneyia* (Wilson 1966), appear to be quite advanced. The "Anthropoidea" hence are not a natural, monophyletic group and cannot be made so unless we include their common tarsioid ancestors, making Anthropoidea equivalent to Haplorhini. But while I have objected, above, to degrading the tarsioids to the level of the "prosimian" hoi-polloi, I do not believe that they should be pushed upward either.

The platyrrhine record is a scanty one. They do not appear in the South American fossil record until the end of the Oligocene. Possibly they may have been present earlier in more tropical regions of the continent; possibly they had only then succeeded in crossing a gap from Central America. Little has been added to our knowledge of them in recent years except for two finds in the Miocene of Colombia (Stirton and Savage 1950; Stirton 1951).

Catarrhines.—The oldest catarrhines, except for problematical specimens from the late Eocene of Burma, are those from the Egyptian Fayûm beds. Here we find a puzzling situation in that in such a form as *Propliopithecus* we already have an animal that is definitely hominoid, whereas other Fayûm materials, such as *Parapithecus* and *Apidium*, are less specialized in nature, and still others are perhaps leading in a cercopithecoid direction. This differentiation at such an early stage has led to the suggestion that the monkey and hominoid lines arose independently from tarsioid ancestors—thus making a threefold, rather than a twofold, split-up of the erstwhile Anthropoidea. However, the sum total of anatomical features common to the Old World monkeys, on the one hand, and to the ape-man stock, on the other, is so great that this does not seem probable. Simons (1959, 1962) is currently reexploring the Fayûm with particular attention to primates, and it is to be hoped that his work will result in clarification of early catarrhine history.

Presumably because of the relative paucity of Tertiary deposits in the more tropical parts of Eurasia and Africa, where the Cercopithecidae have flourished, the fossil finds of this family have not, unfortunately, contributed greatly to the history of this monkey family in recent years.

Oreopithecus from the Italian Pliocene was long dismissed as just another Old World monkey; Hürzeler is to be congratulated for calling attention to the fact that *Oreopithecus* is not a monkey but a hominoid, and even in some features suggestively hominid. His work has resulted in finding more and much better remains of this genus. Although definite allocation should await full publication of the results, it would seem that *Oreopithecus* is definitely a hominoid, but that it probably represents an independent evolutionary line paralleling the advanced great apes and hominids in various features (Hürzeler 1958, 1959; Straus 533).

Over the years numerous remains, nearly all very fragmentary, of manlike apes have been obtained from the Miocene and early Pliocene. It seems clear that the gibbons and the *Pliopithecus* group, to which some of the finds belong, must have separated from the other pongids at an early stage. (Recent additions to knowledge in this area include description of a *Pliopithecus* skeleton by Zapfe [1958], and *Limnopithecus* from East Africa by LeGros Clark and Thomas [1951] and Ferembach [534].) I have not followed some authors who would separate the gibbons at a family rather than a subfamily level. Most of the known pongid remains belong to a more progressive group, among which we would expect forms roughly ancestral to the "higher" anthropoid apes and man. Here we suffer from a surfeit of generic names—about a score of them—for forms which, as far as the evidence goes, differ very little from one another. Simons and Pilbeam (535) point out that most of these might well be bracketed under the old familiar generic name *Dryopithecus*. The East African Miocene specimens, to which the name *Proconsul* was unfortunately attached, can be included in *Dryopithecus* in a broad use of that generic term. Various studies on the relatively adequate materials of "*Proconsul*" have in recent years shed considerable light on ape history (LeGros Clark and Leakey 536; Napier and Davis 1959). We have here a manlike ape, but one of a relatively primitive type, without the excessive brachiating specializations toward which the modern forms trend, and quite possibly close to the ancestry of chimpanzee and gorilla and even man.

Tertiary ape remains pointing in a specifically human direction are few. It seems reasonable, however, as Lewis (1934) first pointed out, to believe that the fragmentary remains known as *Ramapithecus* from India belong in this category as does the similar and perhaps generically identical *Kenyapithecus* from Kenya (Leakey 1962; Simons 1964).

Hominids.—It has for some time been quite clear that the members of the *Australopithecus* group are structurally antecedent to man, and by now, I think, all competent workers, with perhaps one—or two?— exceptions, will agree that, as a group, the australopithecines are phylogenetically antecedent to men as well (Broom and Schepers 540; Robinson 541, etc.). There is still some debate about whether or not several genera are to be distinguished among the australopithecines; I have (improperly, I expect) "lumped" them all, in *Australopithecus* (cf. Robinson 1963, 1966).

A modest degree of progress has been made in knowledge of definitely human types. Here, as in the case of fossil apes, we find a plethora of generic names; certainly in any other group than that of our own close relatives the differences between, for example, "*Pithecanthropus*" and *Homo sapiens* would not be considered as worthy of generic differentiations except in the hands of an extreme "splitter." There is, I think, a current tendency for workers in this field to follow the urgings of such authors as Simpson and Mayr (for example, Mayr 1963) in including all forms from the "*Pithecanthropus*" level on, in the genus *Homo*.

Many workers hold the belief (whether specifically stated or not) that in Paleolithic days man was split geographically into discrete species evolving independently of one another. There is now, due to the urging of Dobzhansky and others (see Dobzhansky 1963), a growing realization that although in earlier times interregional contacts were presumably less than in later ones, there was even then a certain amount of genetic interchange. Certainly in those times, as later and today, there was a trend for regional racial differentiation (cf. Coon 1962), but nevertheless at any given stage of the Pleistocene the human population is to be regarded as having constituted a single, if diversified, species.

Formerly it was thought that members of the "*Pithecanthropus*" = "*Sinanthropus*" stock might have been restricted to the Orient, but discoveries of similar types in North Africa ("*Atlanthropus*" of Arambourg 1963) and East Africa (Leakey 539) have shown that men of the *erectus* group were widespread. Leakey, Tobias, and Napier (1964) have recently proposed the identification in East African materials of a stage below the *erectus* level as *Homo habilis*; at the moment, however, this finding has not been universally accepted (cf., for example, Robinson 1966, 1967).

At the time my second edition was written, Piltdown man was almost universally accepted as an authentic find. The exposure of this hoax a decade ago (Weiner 1955) has removed a major stumbling block in the interpretation of early human history.

19

Carnivorous Mammals

Creodonts.—Matthew in his classic memoir of 1909 (545) adopted a "horizontal" type of classification of carnivorous (or supposedly carnivorous) mammals, grouping in the suborder Creodonta all the older types which might belong or were thought to belong in this general category. As so constituted, the Creodonta were a miscellaneous assembly. The Hyaenodontidae and Oxyaenidae were the prominent creodont families; the Miacidae were clearly ancestral to the fissipedes; the Arctocyonidae and the probably related Mesonychidae were quite different from either typical forms or miacids, and were archaic in many ways.

Recent decades have seen the dissolution of this artificial grouping. Nearly all workers have long since agreed in removing the Miacidae and associating them, in "vertical" fashion, with their fissipede descendants. A more recent trend, as yet little expressed in published literature, is for further reduction of the Creodonta by removal of the Arctocyonidae and Mesonychidae which, as discussed later, seem more appropriately placed in the primitive ungulate order Condylarthra.

This reduces the creodonts to the "core" families, Hyaenodontidae and Oxyaenidae. These two families were the dominant predators of the Paleocene and Eocene. It is generally agreed that the two are closely related and form a natural unit. They do not appear to be at all closely related to the later fissipedes or to the fissipede miacid ancestors and, hence, merit recognition as a major placental group, ordinally distinct from the Carnivora proper.

Their pedigree appears to be a distinctive one, which can be traced back to the Cretaceous separately from that of any of the other well-known placental orders. Simpson (1928) long ago recognized the similarity to typical creodonts of *Deltatheridium* of the late Cretaceous and certain of its Cretaceous and early Tertiary relatives, such as *Palaeoryctes* and *Didelphodus*. There has gradually grown the belief among students of primitive placentals that the deltatheridians are definitely ancestors of the typical creodonts (cf., for example, McKenna 464, p. 86).

It remained for Van Valen (543, 1965*b*) to take the positive step of uniting the deltatheridians and the typical creodonts in a single discrete order. He would also include a few other obscure early Tertiary forms, such as *Didymoconus* [*Tshelkaria*] and *Micropternodus* [*Kentrogomphius*], as the Didymoconidae (the Tshelkariidae of Gromova 1960). This combined group Van Valen would name the order Deltatheridia. The term Creodonta is familiar and available, however, and is most generally understood to be particularly applicable to the hyaenodonts and oxyaenids that form the main substance of the new order. I prefer its use to the unnecessary introduction of a new term.

As discussed earlier, the "zalambdodonts" show dental similarities to the deltatheridians. Part or all of these supposed insectivores might pertain to the order Creodonta, as Van Valen suggests, but relationship to lipotyphlan insectivores appears perhaps more reasonable.

Carnivora.—With the transfer of arctocyonids and mesonychids to the Condylarthra and the erection of a separate order for the typical Creodonta, the Carnivora proper are reduced to the modern families of the Fissipedia and Pinnipedia plus the ancient family Miacidae. Once included in the Creodonta (*sensu lato*) in the horizontal Matthew type of classification of carnivores, the miacids have been recognized, since the days of Cope, to include the ancestors of the modern carnivores; and in recent decades it has been customary to place the miacids among the Fissipedia even when the term Creodonta was retained for the other families usually placed in that group. The general "habitus" of the miacids appears to be continued among "modernized" forms by the viverrids (and to a lesser extent by mustelids), and Gregory and Hellman (552) advocated placing the miacids in the Viverridae. They have not, however, been followed by other workers in this, and it seems preferable to retain the miacids in a separate, ancestral infra-order or superfamily of Fissipedia, of equal rank with arctoid and aeluroid groups. Little recent work has been done on miacids, despite the evolutionary importance; it is hoped that a very recent publication by MacIntyre (1966) on two miacid genera will be followed by others to give us a modern study of the family as a whole.

The origin of the family—and hence of the Carnivora as a whole—is far from clear. It has been variously claimed that they are close to typical creodonts or to arctocyonids, but there is no strong evidence. MacIntyre discusses this subject at some length but fails to reach any positive conclusion. At present an independent origin from primitive insectivores is as reasonable a conclusion as any.

The end of the Eocene and beginning of the Oligocene, including the time of deposition of much of the Quercy phosphorites, is a cloudy period in fissipede history, on which more light—difficult to obtain—is needed. This was a time when the miacids were evolving into higher fissipedes, and it is difficult to distinguish late miacids from early arctoids and aeluroids, as such good workers as Schlosser and Teilhard de Chardin found; further, it appears probable that a split between the two later groups had already occurred within the limits of the miacids themselves. The miacid-viverrid transition is so slight that, as mentioned, Gregory and Hellman advocated including the miacids in the Viverridae, and although later mustelids are readily distinguishable, some of the early types are similarly difficult to sort out.

Due to the tropical forest-dwelling habitat of the viverrids we still have little fossil data shedding light on their history; due to their occurrence in more temperate climatic zones, data on mustelid evolution is much more plentiful despite their generally forest-dwelling habitat. The rapid differentiation of the felids is still a startling phenomenon. Since I have not concerned myself with subfamilies, I have shied away from the problem, as yet not resolved to everyone's satisfaction, of the relation between "normal" felid types and the various series with enlarged sabers; in addition to phylogenetic difficulties, the systematics of the situation have not been eased (to say the least) by a considerable accretion of generic terms in recent decades. Being a mild "lumper" by nature, I have included all the living felids (except *Acinonyx*, the cheetah) in *Felis*. Simpson in 1945 "lumped" a majority of the two dozen or more supposed cat genera in *Felis*, but allowed half a dozen to be placed in a separate genus *Panthera*. I see few features here, however, to warrant generic separation except those simply associated with larger size.

The Canidae as presently constituted are a large and highly varied group; I have reduced them to some slight extent by shifting the "*Cynodon*"-*Hemicyon*-*Cephalogale* group to the Ursidae, to which they probably gave rise. Several authors have advocated further cleavage, but perhaps the sorting of subfamily assemblages (with which I have not concerned myself) is sufficient. In middle and late Tertiary times, dental characters appear to be in general sufficient for family assignment. But, as Hough has shown (551, 1944), this is not always the case when skull and ear structures are investigated; *Phlaocyon* and *Aletocyon*, for example, appear dentally to be primitive procyonids but are actually canids; *Zodiolestes* has a mustelid dentition but a procyonid ear structure. The Procyonidae are poorly known as fossils (perhaps in relation to forest dwelling). *Ailurus*, the "true" small panda of Asia appears to be, in contrast to the "giant panda," a proper procyonid.

Opinions have varied greatly over where to draw the limits of the family Ursidae. Apart from the distinctive forms of the Pleistocene and Recent, all agree in including *Ursavus* and *Agriotherium* of the late Tertiary, but although it seems fairly clear that the bear pedigree traces back to the *Cephalogale*-*Hemicyon*-"*Cynodon*" group, these forms are frequently retained as a subfamily among the Canidae. Erdbrink (562), in a thorough study, advocates including these presumably ancestral forms in the Ursidae, and I have followed him. He further points out that supposed generic distinctions among the modern bear genera are, for the most part, of trifling value and concludes that *Thalarctos*, *Helarctos*, *Selenarctos*, and *Tremarctos* should all be included in *Ursus*; however, Thenius (1959) and Kurtén (1966) consider *Tremarctos* and *Melursus* as valid.

The position of *Ailuropoda*, the "giant panda" of western China, has long been a matter of controversy, some believing it allied (through *Ailurus*) to the procyonids, others suggesting ursid relationships. As the result of careful detailed study, Davis (567) claims that this interesting form is definitely ursid, but doubts persist.

A number of Tertiary faunal studies of carnivores have been published in the last two decades. We may note, for example, those of Thenius on Göriach carnivores (1949), of Dehm (1950) on Eichstätt, Viret (1951) on Grive Saint-Alban, and of Ginsburg (1961) on Sansan. Kretzoi (1945) proposed a radical reclassification of carnivores which has not, however, been well received by other workers.

Pinnipeds.—Little new data have come to light on the history or origin of the Pinnipedia in recent decades. It is agreed that the eared seals (Otariidae) and walruses (Odobenidae) are related to one another and are to be contrasted to the "true" seals of the family Phocidae; various authors, of whom McLaren (1960) has been the most recent advocate of the thesis, have suggested a dual origin for pinnipeds. A dual origin is unproved but not out of the question; the potentiality toward parallelism in an aquatic trend among aeluroids is shown not only by the evolution of the marine otter but also by the interesting Pliocene *Semantor* (Orlov 1933), thought at first to be a special type of pinniped but now believed to be a specialized mustelid (Thenius 1949).

20

Archaic Ungulate Groups

Condylarths.—This order has been considered, with reason, a basal ungulate group, from which part and possibly all other ungulate stocks may have been derived. Based primarily on *Phenacodus*, the content of the order has varied from time to time and from author to author. In addition to the Phenacodontidae, generally included components are the Hyopsodontidae (with the *Mioclaenus* group), the Didolodontidae as a series of South American survivors, and probably (despite somewhat conflicting opinions by one author or another) the Periptychidae and Meniscotheriidae. Except for the didolodonts, members of families customarily placed in the order are almost exclusively Paleocene and North American; their restriction in space is very probably due to lack of adequate knowledge of the Old World Paleocene. Dehm and Oettingen-Spielberg (467) have recently described a hyopsodont from Pakistan. Lemoine long ago described three possible condylarths from the late Paleocene of France; these have been restudied by Russell (461) and are considered members of the order. *Phenacolophus* of the late Paleocene of Asia, doubtfully assigned here by Matthew and Granger (1925), is perhaps best considered as an insectivore of doubtful affinities. The didolodonts, whose condylarth affinities were first recognized by Ameghino, are for the most part very early Tertiary. McKenna (1956), however, has described a late survivor, *Megadolodus*, in the Miocene of Colombia. He points out that *Lophiodolodus* of the Oligocene is not a didolodont, whereas *Protheosodon* does pertain (points which I regret to have overlooked in my generic listing).

In my present text I depart radically from customary classifications in including in the Condylarthra two families, the Arctocyonidae (Oxyclaenidae) and Mesonychidae, which are generally placed in the Carnivora. In this I am following a line of thought which has gradually developed among workers on early Tertiary mammals. Although assigned by Matthew and other workers of earlier decades to the Creodonta, the arctocyonids show no particular resemblance to the more typical creodonts of the hyaenodont or oxyaenid families, or to the fissipedes. On the other hand, there is little to distinguish arctocyonids from certain condylarths. Simpson (431, p. 234), for example, noted this; and this similarity may have been in great measure responsible for his grouping the carnivores and ungulates together as Ferungulata. Patterson (438, p. 18; and in Patterson and McGrew 1962) suggested assignment of the family to the Condylarthra, and McKenna (in Russell and McKenna 1961) noted arctocyonid-condylarth resemblances. There is no indication of any sort of genetic relationship of arctocyonids to "typical" creodonts or true carnivores. The same argument in general applies to the Mesonychidae. There is no positive reason for inclusion of these blunt-toothed, hoofed creatures among the creodonts or true carnivores; their relationship to the arctocyonids (and through them to the condylarths) seems reasonably clear.

The condylarths are certainly a primitive group of ungulates. Are they the ancestors of all ungulates? It has long been felt that the perissodactyls are direct descendants of the phenacodonts, and the difference between such a phenacodont as *Tetraclaenodon* and *Hyracotherium* is slight indeed (Radinsky, 1966*b*). The didolodonts, again, seem clearly to lead to the Litopterna. Beyond this, the evidence is less clear. Simpson (431), among others, has noted resemblances between the Hyopsodontidae and artiodactyls, but a fair interordinal gap remains. The same author has grouped the Notoungulata and Astrapotheria with the Condylarthra and Litopterna in a common superorder Protungulata but cites no evidence for the "reasonable, even though doubtful conclusion" that the notoungulates are of condylarth origin. Quite possibly all ungulates are of common descent; but, as stated earlier, it is equally possible that more than one group of insectivore descendants took to a vegetarian terrestrial life and attained ungulate status.

Gazin (1965) has recently restudied *Meniscotherium*. He does not believe that *Meniscotherium* has descendants; hyracoid resemblances are due to convergence. Dehm and Oettingen-Spielberg

(467) have recently described new mesonychids from Pakistan. Of great interest is the identification by Sloan and Van Valen (451) of a late Cretaceous pre-ungulate, *Protungulatum*, which they would place in the Arctocyonidae.

Pantodonts.—The Pantodonta were the main constituents of Cope's Amblypoda with *Coryphodon* of the Lower Eocene as essentially the type form. Since, however, the uintatheres were generally included in the Amblypoda but are now generally considered to be a quite distinct group, most recent authors have, to prevent possible ambiguity, followed Simpson's lead in using Pantodonta for *Coryphodon* and its allies.

Patterson's work in the 1930's called attention to the presence, in the late Paleocene, of impressive forms belonging to the pantodont group but differing considerably from *Coryphodon*; in consequence Simpson in his 1945 classification included not only the family Coryphodontidae but a second family, the Barylambdidae, for these new forms. Since that time further additions to the pantodont assemblage have been made both in North America and Asia. The group was reviewed by Simons in 1960 (576). He suggested the erection of two additional families, separating the Pantolambdidae from the Coryphodontidae, and considering *Titanoides* (including *Sparactolambda*) distinct enough to merit family separation. Flerov (1952) advocated a separate family for the Asiatic *Archaeolambda*; Simons, however, believes this form to be generically identical with the pantolambdid *Haplolambda*. *Pantolambdodon*, known only from lower-jaw material from the Asiatic Upper Eocene, is considered by all recent workers to be of dubious status, and not surely even pantodontid.

Uintatheres.—These ugly horned creatures whose characteristic members are most common in the Middle Eocene Bridger beds (and whose description entered prominently into the early phases of the Cope-Marsh feud) were long considered to form, with the pantodonts, a single order Amblypoda. Due, however, to the studies in the 1930's of Simpson (572) and Patterson (574), it became clear that the similarities were mainly, at least, habitus ones, and that clear distinction of the two was required, with the use of the Marsh term Dinocerata for the uintatheres.

The uintatheres were formerly arrayed in a single family. But with increasing knowledge of pre-Bridger forms and Asiatic types, Flerov (1957) reasonably suggested a division into three families: Prodinoceratidae for early forms, Uintatheriidae for the typical Bridger and late Eocene genera, and Gobiatheriidae for the peculiar late Eocene Mongolian *Gobiatherium*, whose relationships with other members of the group are unclear. Dorr (1958) has discussed uintathere phylogeny. Wheeler (575) has published a revision which is particularly useful in attempting to sort out the taxonomic tangle of the plethora of Cope-Marsh names.

The earliest uintatheres had already developed, as Patterson noted, the unusual molar pattern and other characteristics of the group as a whole. We have no clue as to the earlier Paleocene history of the uintatheres or their ancestors.

Pyrotheres.—No recent light has been shed on *Pyrotherium* of the South American early Oligocene, with its peculiar proboscidean-like adaptations. It stands quite isolated, except for a few other early South American forms which are very poorly known or of doubtful pertinence. (As noted below, the Paleocene *Carodnia* is not ancestral.) I note the suggestion by Sera (1954) that *Pyrotherium* was an aquatic proboscidean; this would account for its presence in South America but is not otherwise helpful.

Xenungulata.—Simpson had described some very fragmentary remains from the Paleocene of Patagonia as *Carodnia* (and *Ctalecarodnia*); Paula Couto (460) found further remains, including a partial skeleton, in the Paleocene of Itaborai, Brazil. There are bilophodont molars, due to which Simpson suggested a relationship to *Pyrotherium*; Paula Couto, however, found in his better materials so many differences that he proposed a new order Xenungulata for this form. *Carodnia* left no known descendants, and no definite proof of relationship exists, although Paula Couto notes points of resemblance to uintatheres.

Amblypoda.—We have discussed briefly, above, several types of early ungulates, mainly from the Americas, of clumsy archaic build with low-crowned lophodont molars; further forms of the same sort (most prominently the Proboscidea) appear to have originated in Africa. Each of these types is generally given ordinal rank. Simpson (431) has suggested uniting these archaic ungulates in a superorder Paenungulata, although admitting the strong probability of polyphyletic origin of these forms from primitive ungulate or pre-ungulate ancestors. The nucleus of his proposal is, he states, the familiar association of proboscideans and the other African types, commonly referred to as "subungulates." That these forms are a natural unit is not improbable, but I hesitate to associate them with the four essentially American groups just discussed, whose common origin (if any) may well have been independent of that of the subungulates. The term Amblypoda was long employed to include pantodonts and uintatheres. Wheeler (575) has suggested the revival of this term to include these two groups and *Carodnia* as well, and in my text I have broadened this suggestion to include the pyrotheres as a further orphan group of "amblypods." Quite possibly all four are purely parallel in origin; but I think their association in a common group is at least as justified as Simpson's broader fusion of them with the subungulates in his Paenungulata.

21

Subungulates

Proboscidea.—Often referred to, since the days of Gill a century ago, as subungulates is an assemblage of primitive ungulates or forms which are believed to be of ungulate descent and which are probably of African origin. These include (1) the Hyracoidea, (2) Proboscidea, (3) Embrithopoda, (4) Sirenia, and (5) the recently recognized order Desmostylia.

Most prominent of these groups is the Proboscidea. The literature on this group is enormous. All the older works were reviewed in Osborn's (580) mammoth work on the order; scores of publications on fossil proboscideans have appeared in recent years, but few are of major importance. Osborn's work is difficult to deal with, particularly as regards nomenclature. In his 1945 classification, Simpson gave a welcome simplification of the nomenclature which I have in general accepted with pleasure. I have even followed him, although dubiously and with distaste, in using *Gomphotherium* instead of the better and frequently used term *Trilophodon*. I have, however, balked at calling the common *Mastodon*, "*Mammut*." Since the date of Simpson's revision, there has been a continuation of the proposing of new and unnecessary generic names for proboscideans; I have in general ignored them.

Owing to Osborn's peculiar and individual ideas on evolutionary processes, hardly a single proboscidean is considered by him (as can be seen from his phylogenetic chart) to be ancestral to any other. As Simpson points out, in contrast to Osborn's definitive opinion, the rather generally accepted belief that the elephants are of stegodont ancestry, and that the stegodonts in turn are of mastodont derivation, is a reasonable one.

Deinotherium, lacking upper tusks, but with curiously downturned lower tusks and a very primitive molar build, appears suddenly in Eurasia and Africa in the Miocene and survives little changed into the Pliocene in the north and into the Pleistocene in Africa. We have no clue as to its earlier evolutionary history, perhaps because of the paucity of African Oligocene deposits later than the Fayûm beds. It is always included in the Proboscidea, although placed in a separate suborder. There is, however, little positive evidence, as far as I am aware, to show that *Deinotherium* is actually a member of that order rather than a parallel evolutionary development from a primitive subungulate stock.

Barytherium, from the Upper Eocene of the Fayûm, is known only from teeth, a lower jaw, and a few limb bones. No new data have come to light since its original description by Andrews half a century ago. Its presence in Africa and its molars with their transverse lophs suggest, reasonably, that it is a subungulate of some sort; but there is no evidence that it is related particularly to any other subungulate type. It seems absurd (despite the analogy with *Carodnia*) to erect an independent order for such poor remains, and I have, therefore, like others before me, tucked it away in a discrete suborder of the Proboscidea.

Embrithopoda.—One of the most spectacular of fossil mammals, the huge-horned *Arsinoitherium* stands completely isolated. As in the case of *Barytherium*, no new data have come to light since Andrews' description fifty years ago. Here again, we can only assume that we are dealing with a specialized offshoot of the ancestral African subungulate stock.

Sirenians.—Only a modest amount of data has been added to our knowledge of fossil sirenians in the last two decades. There is now a rather full documentation of the history of the dugong family. But we still know very little of the history of the manatees. I refuse to follow the absurd dictate of the nomenclature commission that *Manatus* be outlawed and that these mild creatures have forced upon them the name *Trichechus*, which from the days of Linnaeus himself had been almost universally applied to the walrus!

Desmostylia.—For many years, our knowledge of *Desmostylus* and its relatives—*Cornwallius*, *Paleoparadoxia*, and *Vander-*

hoofius—was in the main confined to their peculiar teeth, present in deposits of the Miocene and Pliocene on both sides of the North Pacific. The only important exception, until recently, was a single skull (which Abel in 1928, curiously, claimed to be that of a monotreme). Since desmostylian teeth were found only in nearshore marine deposits, and since the teeth show some slight resemblance to those of sirenians, it has been until recently assumed that these forms were a side branch of the sirenian stock. Reinhart (582) a few years ago reviewed the then existing data on the group which, contrary to most other workers (except Sickenberg 584), he advocated elevating to ordinal status as the Desmostylia.

That we are dealing with a separate order is now clear from the startling discovery that the desmostylians were not stream-lined aquatic forms but had a set of four stout limbs and a somewhat hippopotamus-like build! A desmostylian skeleton was discovered in Japan in 1934, and the skeleton prepared and mounted half a dozen years later. But although photographs of the skeleton were distributed a number of years back, it was not until 1966 that a printed description was issued (Shikama 1966). Meanwhile, in North America, the excavation for an electron accelerator on the campus of Stanford University un-covered another skeleton, which I have figured in my text (Fig. 367) from a preliminary report by Repenning.

The desmostylians, thus, were not specialized swimmers but browsers (on water plants?) along the coasts. Despite the contrasts with the sirenians, Shikama finds no reason to depart from the assumption that the desmostylians are a subungulate stock. But how and when they reached their circumscribed North Pacific home is a mystery.

Hyracoids.—Concerning the coney group, somewhat removed from the other subungulates but generally accepted as a part of this complex, a few additional finds have come to light in recent years. As might be expected, Leakey's work in East Africa has brought new material to light (Whitworth 1954); and hyracoid remains have been discovered in the Pontian of France (Viret 1949). The group has been relieved of a dubious appendage by Patterson's (518) pointing out that *Myohyrax* and *Protypotheroides* are not hyracoids at all but elephant-shrew relatives. Whitworth suggests that the conies should be ranged in a group of "Mesaxonia" with the perissodactyl ungulates; needed, however, is (as Simpson has pointed out) a thorough comparative study of hyracoid anatomy.

Subungulate interrelationships.—The assumption of the unity of the subungulates is founded in great measure upon the geographical picture. The dearth of information regarding fossil mammals in Africa for most of the Eocene as well as the Paleocene is lamentable. How much evidence regarding the relationships of the subungulate groups can be gained from anatomical studies? As far as I am aware, no one has considered this matter at all deeply since Matsumoto (579) forty years ago. A thorough review not only of osteology but also of the soft anatomy of the surviving subungulate groups might be of great value.

Moeritherium is universally regarded as a component of the Proboscidea. Consideration of a skeleton of this creature, collected by Simons and mounted in the Yale University Museum, gives one pause, however. With its long low-slung body it has a general appearance quite different from that of a typical proboscidean, and suggests marsh-dwelling, somewhat amphibious habits. Is *Moeritherium* specifically a proboscidean ancestor? Is it not, perhaps, representative of a group of a more primitive nature, close to the common ancestry not only of the mastodons but also of the sirenians and perhaps other subungulate types?

22

South American Ungulates

The greater part of the fabulous ungulate fauna of South America was originally collected and described by the remarkable team of Carlos and Florentino Ameghino. Good work by Argentinian scientists has continued, but much collection and description has been done by North Americans. Hatcher's great Patagonian collections at the turn of the century were described by Scott and Sinclair; Loomis collected and described fossils from the *Pyrotherium* beds; Riggs' Middle Tertiary collections in the 20's were described in great measure by Patterson; in the 30's Simpson collected and published, with effect, on Lower Tertiary faunas. In 1948 Simpson (459) reviewed the earlier phases in the history of part of the South American ungulate groups. Relatively little of importance has been published since that date, although Stirton's work in Colombia has added some geographic breadth to our knowledge. A completion of Simpson's review is in press.

Litopterna.—Although some early writers tended to bracket the litopterns with the notoungulates, all modern workers agree that they are a distinct order. Further, it is very generally agreed that they are definitely condylarth descendants (Simpson 431); the litopterns are in a sense merely a continuation of the condylarth group, and it is difficult to refrain from grouping them in a common order. The presence of the didolodontid condylarths in the earlier Tertiary of South America and the condylarth-like nature of some of the earliest litopterns tend to close the gap (Paula Couto 460; McKenna 1956). Except for the description by Paula Couto of further Paleocene materials, almost no new data have been added in recent years to our knowledge of litopterns. There are two universally recognized families, Proterotheriidae and Macraucheniidae; as some earlier authors have done, I have recognized the Adianthidae as a family distinct from the Macraucheniidae.

Notoungulates.—By some authors all South American ungulates were included in a single order, but due to the work of Scott, Simpson, and Patterson, among others, it seems now clear that, as Roth saw more than half a century ago, litopterns, astrapotheres, and pyrotheres all belong to groups discrete from the main body of ungulates of that continent, presently termed the order Notoungulata. The classification of the forms included in the Notoungulata is difficult, partly because of the vast number concerned and further because of the great mass of genera and species named by Ameghino on the basis of poor material. Many may be clustered in suborders centering around two well-known Pleistocene forms *Toxodon* and *Typotherium* (which the priority-seekers insist must be called *Mesotherium*); *Homalodotherium* was long considered representative of a third suborder Entelonychia. In 1945 Simpson proposed radical changes. Consequent on Patterson's work on *Homalodotherium*, the Entelonychia were fused with the Toxodontia, whereas *Hegetotherium* and its allies, which superficially resemble the typotheres, appear to represent a discrete stock and were placed in a separate suborder. Important was Simpson's segregation of a series of primitive types from the Paleocene and Eocene to constitute a basal suborder Notioprogonia.

I have in my classification followed Simpson's system for the most part. As he notes, the proper position for the Oldfieldthomasiidae, Archaeopithecidae, and Archaeohyracidae is open to doubt. I have retained the first two in the Toxodontia, where Simpson placed them, but transferred the Archaeohyracidae to the Hegetotheroidea. Some further systematic revision is expected in Simpson's work now in press.

In early decades of the century, startling discoveries were the finding of abundant specimens of a primitive notoungulate in the Paleocene of Mongolia and of a single specimen in the Wyoming Eocene. Except for an unpublished North American Paleocene specimen, nothing further has since been reported of notoungulates outside the confines of South America and, apart from the reasonable but unproved assumption of condylarth derivation, we know nothing of when, where, and whence this great vanished order was derived.

Astrapotheres.—As regards the Astrapotheria, once included by some writers in the notoungulates but now generally considered an independent order, little new data have been added in recent years. Simpson, in his publication now in press, expresses the belief that *Trigonostylops* (now known from the Paleocene as well as Eocene) is not a member of the astrapothere group and should be placed in a separate order.

23

Perissodactyls

In contrast to some mammalian orders with vaguely defined boundaries and imperfectly defined internal divisions, the Perissodactyla rejoice the heart of those who like neatness and precision. Before the evolution of the group had proceeded far in time, ancestors of four advanced types can be made out, leading to the equids, titanotheres, chalicotheres, and rhinoceroses. Within each of these groups is a wealth of material, and the greater part of the evolutionary story is rather adequately known. It is only at the base of the phyletic story that some possibilities of confusion exist. The order appears in the Lower Eocene, without a definite Paleocene ancestry (except for a single late Paleocene specimen, as yet undescribed). There is, however, very general agreement that the condylarths are ancestral and the ordinal gap is slight; Radinsky (1966*b*) has argued that the phenacodontid genus *Tetraclaenodon* is almost exactly the type to be expected in the condylarth ancestor. Before the termination of the Eocene the members of the advanced groups are almost all clearly distinguishable; but early in that period there is a situation in which, as one would expect, early members of the different lines are difficult to sort out, and we are dealing with a somewhat "bushy" evolutionary growth; were there no later representatives, it would not be unreasonable to include most of the early forms in a single family (or superfamily, at the most). "*Eohippus*" is traditionally placed at the base of the horse family, but it is generally agreed that it is very close to the ordinal base and has little to show its specific antecedence to equids rather than to other lineages.

The modern tapirs are, of course, the least modified of perissodactyls, and hence there is general agreement that early forms not definitely assignable to horses, titanotheres, chalicotheres, or rhinoceroses be grouped with these modern forms as tapiroids. Just which early forms should be included in this assemblage and how arranged in families has been uncertain. Two decades ago, both Simpson and I arrayed the tapiroids in four families—Helaletidae, Lophiodontidae, Isectolophidae, and Tapiridae—basing this mainly on the work and advice of H. E. Wood. In recent years Radinsky has begun a thorough study of early perissodactyls; the results of his early work, including a major paper on North American tapiroids (605) were utilized by me for the present edition, and he later published a study of early Asiatic forms (606). Radinsky further subdivided the group, making separate families for the Lophialetidae and Deperetellidae. Still further, the hyrachyid Eocene genera, earlier studied by Wood (609) and generally considered as rhinoceros relatives, are thought by Radinsky to be "generalized" tapiroids rather than rhinocerotoids, to be bracketed with helaletids.

The major evolutionary story of the horses has long been known, a wealth of American material substantiating the brilliant early deductions of Kowalevsky. Once believed to be an example of orthogenesis, we now know that the equid story, like that of many other mammalian evolutionary series, shows many branches, from the time of the divergence of lophiodonts in the Eocene to a late Tertiary radiation which is still not fully clear. Quinn's (1955) study of late Tertiary forms from the Gulf Coast, in which he believed that a considerable number of parallel lines could be sorted out among the forms generally lumped in the genus *Merychippus*, has not met with general acceptance; but it has the great merit of pointing out the existence here of an unsolved problem. Of great general interest is Edinger's (595) study of the horse brain, demonstrating parallelism in the development of advanced mammalian brain types from a primitive eutherian stage.

The curious clawed chalicotheres, once denied admission to the order, have long been thought to be includable in a common suborder Hippomorpha with the horses and titanotheres, and in 1945 I took (apparently inadvisedly) the further step of placing them, as suggested by the dentition, in a common superfamily with the titanotheres. Radinsky's studies, however, incline him to an opposite conclusion—that the chalicotheres should be separated from both horses and titanotheres and placed in a discrete suborder for which the old term Ancylopoda may be revived.

If we accept Radinsky's conclusion that the hyrachyids be removed, the rhinocerotoid superfamily then includes the sprightly running rhinoceroses, the Hyracodontidae of the Eocene and Oligocene, the heavily built and essentially contemporary Amynodontidae, and the "true" rhinoceroses, the Rhinocerotidae. This last family has had a long and exceedingly complex history; Radinsky (1966*a*), however, would remove from the rhinocerotid family a number of early forms and restrict the Rhinocerotidae to genera in which the typical "modern" incisor pattern was developed. H. E. Wood devoted the greater part of his scientific career to a study of this group. His results have been published in part in various papers, but it is a source of great regret that illness will presumably prevent him from giving a final summary of his mature opinions and conclusions.

24

Artiodactyls

In contrast to the perissodactyls, whose components can be for the most part tied up into neat packets, the artiodactyls, when fossil forms are included, are a vast sprawling order among whose components relationships and evolutionary lines are all too commonly vague and uncertain. The living forms can be sorted out fairly neatly into (1) pigs and peccaries, (2) hippopotomi, (3) camels, (4) chevrotains, and (5) advanced ruminants or pecorans, including deer, giraffes, bovids, and prongbuck. All five may be given subordinal rank; or, as is sometimes done, 1 + 2, which are bunodont as to dentition and are not cud-chewers, may be lumped as non-Ruminantia, Suina, or Hyodonta, while, in contrast, 3 + 4 + 5, which have a selenodont dentition and are cud-chewers may be designated as Ruminantia or Selenodontia. The camels are clearly distinct from other living selenodonts and are generally given separate subordinal rank as the Tylopoda, however. Further, although the little chevrotains are clearly structurally (if not directly phylogenetically) antecedent to the "higher" ruminants, many would preserve them as a distinct suborder, the Tragulina, in contrast to the proper Pecora.

We may thus, as regards living families, exhibit some diversity in formal classification, but the general picture is clear. Far different is the situation when, in addition to the existing eight or nine families, nearly a score of extinct groups are injected into the picture. I do not feel too happy or certain about any type of recent classification (my own attempts included).

The first of a series of "modernized" classifications appeared in 1929 by Matthew (618), who in his day had the broadest knowledge of Tertiary placentals of any worker, and whose opinions were highly worthy of consideration. Matthew abandoned the primary cleavage into ruminants and non-ruminants and, instead, divided the order into five suborders. The surviving forms and their obvious fossil relatives comprised three of the five: Hyodonta included the pigs and hippopotami, while instead of lumping all ruminants into a single suborder, he separated the Tylopoda, for the camelids, from the Pecora, in which last he included the tragulines and a number of moderately advanced American forms such as *Leptomeryx*, *Hypertragulus*, and *Protoceras*. Two completely extinct suborders were added. He (properly, I think) erected a suborder Protodonta for truly primitive artiodactyls often included in the Dichobunidae (*sensu lato*), together with several other early and relatively primitive types, such as the "giant hogs"—the entelodonts or elotheres. Then there are a number of families, selenodont and perhaps ruminating in habits, that are not up to the pecoran level nor, according to Matthew, related to the camels. For this considerable assemblage, in which major components were the oreodonts, anthracotheres, and anoplotheres, Matthew coined the subordinal term Ancodonta.

Scott, in his 1937 version of *Land Mammals in the Western Hemisphere* (437) and in the 1941 artiodactyl section of his revision of the White River fauna (469), followed Matthew in considerable measure. He reverted to the older fashion of first cleaving the order into non-ruminants and ruminants, however, with the first including Palaeodonta, Suina (Matthew's Hyodonta), and Ancodonta, the ruminants including the Tylopoda as well as the Pecora and the Tragulina which (as done by various earlier authors) he separates from the "proper" Pecora. Further, Scott gained the concept of a primitive selenodont assemblage, of which camels are the modern representatives. Accordingly, he reduced Matthew's Ancodonta, as far as forms present in America are concerned, to the anthracotheres and included in the Tylopoda, in a broad sense (with some hesitation), the oreodonts and their agriochoerid relatives and further the hypertragulids, which Matthew had placed in the Pecora.

Simpson (431) in 1945 tried a different approach which, like Scott's, is a partial reversion to classifications of the pre-Matthew type. The camels, with the somewhat parallel xiphodonts attached to them, are left in an intermediate suborder

Tylopoda; all other artiodactyls are placed in the two "classic" groups of Suiformes (non-Ruminantia) and Ruminantia. The latter are divided into tragulines, with which the *Hypertragulus-Leptomeryx-Protoceras* groups are incorporated, and Pecora. Apart from the tylopods and ruminants, all artiodactyls are pushed into the suborder Suina. We find here such varied types as dichobunids and other primitive forms, the pigs and peccaries proper, the hippopotami and their anthracothere predecessors, and a variety of relatively primitive selenodonts, of which the oreodonts are the most prominent. Within the suborder Suina, the discrete components are reasonably treated as the infraorders Palaeodonta, Suina, Ancodonta (with anoplotheres, anthracotheres, hippopotami, and cainotheres), and, as a near novelty, an infraorder Oreodonta was erected for the oreodonts and agriochoeres.

In the Piveteau Traité the Artiodactyla were reviewed in 1961 by Viret. He essentially follows Simpson in his major categories but prefers "Bunoselenodontia" to "Ancodonta" and, further, would include here the xiphodonts, which Simpson bracketed with the camels.

I feel that the Simpson-Viret system of "lumping" a host of diversified forms into a suborder Suina is a distinct retrograde step—rather comparable to terming all non-vertebrate animals "Invertebrata." By doing this one sidesteps the difficult problem of classifying a variety of artiodactyls below the advanced ruminant or camelid level, but does not aid in a solution of the problem concerned.

If one looks broadly at artiodactyl history, one sees a restricted number of truly primitive forms, or of descendants of such forms which have not progressed far, to which Matthew's term Palaeodonta is reasonably applicable. Beyond this base of early Tertiary (mainly Eocene) forms, there seems to be a complex radiation in two general directions: one in which there is a retention of a simple, more or less bunodont, tooth pattern, relatively primitive locomotor apparatus, very probably a mixed diet, and presumably no rumination; and a second in which there is a strong trend toward selenodonty, probably a purely vegetarian diet, perhaps a trend toward rumination and, on the whole, an improved limb pattern. Apart from primitive forms, then, we see a trend toward a cleavage into two groups, for which the old terminologies of Suina or non-Ruminantia and Ruminantia or Selenodontia seem appropriate. I attempted (with some hesitation) to express this point of view in the classification given in my second edition by dividing the artiodactyls into three suborders—Palaeodonta, Suina, and Ruminantia. In Suina I included, as well as pigs and peccaries, some of the more piglike components of the Palaeodonta of other authors and the anthracotheres. In the Ruminantia, using that term in a broad sense, I included all the selenodont families, fossil and Recent. The pecorans, in the usual restricted use of that term, plus the tragulids seem, on present evidence, to be a natural monophyletic group; I therefore considered them as forming an infraorder Pecora (*sensu lato*). There remain, of selenodonts, a number of families—cainotheres, anoplotheres xiphodonts, oreodonts, agriochoeres, and camels—which represent a number of parallel (but perhaps related) essays in advance toward the true ruminant condition. The living camels represent this stage, and the term Tylopoda seemed appropriate to apply to this assemblage of more primitive ruminant families. With some slight modifications I have retained this classification in the third edition.

Numerous objections may, of course, be raised against this attempt at classification. It is highly probable that the early pattern of artiodactyl phylogeny was bushlike, with various slender and distinct "sprouts" rather than compact main branches, so that, for example, a number of the piglike bunodont forms which I have included in the Suina may represent parallel lines rising quite discretely from a primitive artiodactyl base. The Suina, as I use the term, may therefore be far from monophyletic. But it is certainly a closer attempt at a "natural" classification than the Suina of Simpson and Viret, in which are included types ranging from bunodont pigs to such advanced selenodonts as cainotheres and xiphodonts. And if nearly all artiodactyls except pecorans and their kin are to be "lumped" in the Suina, why should not the camels be included in this miscellaneous packet as well?

Just as may be the case with the piglike forms, may not the various selenodont stocks quite possibly represent a series of lines arising separately from the very base of the artiodactyl order? But the excellent studies of Eocene North American faunas by Gazin (465, 466, 631) in recent years have tended in part—although only in part—to strengthen the conclusions expressed in my 1945 classification. Gazin definitely associates the oreodonts (plus agriochoerids) with the camelids, as Scott and I had done, and in contrast to Simpson. These are the major American components of the Tylopoda in the broad sense in which I would use the term. Gazin was not, of course, concerned with the European groups (cainotheres, anoplotheres, xiphodonts, amphimerycids) which I have also included in a broadly defined Tylopoda. We differ, however, in that Gazin would also include in this assemblage *Leptomeryx* and its allies, and the *Protoceras* group which he would (in contrast to general practice) separate from the Hypertragulidae.

The most primitive of known artiodactyls are those mainly described by Sinclair (620) from the Lower and Middle Eocene of North America. Certainly directly descended from them are a number of late Eocene forms and European Oligocene genera such as *Dichobune*. All are frequently included in a single family Dichobunidae, which covers a wide spread between truly primitive forms and others, such as *Dichobune* itself, already rather advanced and specialized. In 1945 I suggested a division into Homacodontidae and Dichobunidae, and in the present text edition I advocate a further separation of the most primitive forms, such as the Diacodectidae (Gazin gives them subfamily status). *Achaenodon* is often associated (as, for example, by Scott and Simpson) with the entelodonts; Gazin, however, points out its relationship to *Helohyus* and would place these forms in the Homacodontidae as a subfamily. I have compromised by giving them family status in the palaeodonts in association with the other "dichobunid" components. Closely allied, too, appear to be the little group of leptochoerids (Macdonald 621).

Beyond this point, what should be included in the Palaeodonta is problematical. Generally included here are "giant hogs," the entelodonts (elotheres). They are obviously a fairly early offshoot from near the base of the artiodactyls and are certainly not closely related to the pigs; on the other hand, they certainly parallel the suoids and, following Pearson (and in contrast to earlier writers who tend to associate them with the dichobunids because of hypocone development), I have included them in the

Suina, as well as the also piglike Cebochoeridae and Choeropotamidae.

In attempting, as I have, to cleave early artiodactyls into "pig" and selenodont lines, the anthracotheres create the most embarrassment. Their primitive representatives are rather similar to the cebochoerids, which seem to be swine, and if, as Colbert particularly (481, 626) maintains, the hippopotami are anthracothere derivatives, we are dealing with the ancestors of bunodonts. Yet the anthracotheres themselves tend toward a selenodont tooth pattern suggestive of relationships to other "bunoselenodonts." I have placed the anthracotheres and hippopotami in a separate superfamily Hippopotamoidea in the suborder Suina; not improbably, however, they may represent an evolutionary line from palaeodont ancestors distinct from either suoids or selenodonts.

As noted above, Gazin's work tends to confirm the unity of the American groups which I have united in a primitive "ruminant" infraorder Tylopoda. As regards details, I have followed Gazin in distinguishing a separate family Oromerycidae for *Protylopus* and a few other early cameloids. The oreodonts have undergone thorough systematic study and revision by Schultz and Falkenbach (632) in a series of studies now near completion. Concerning the European groups which I have included in the Tylopoda (*sensu lato*)—cainotheres, anoplotheres, xiphodonts, amphimerycids—I am not aware of any detailed studies since Hürzeler's major work on cainotheres (633). The cainotheres are, as shown by their distinctive molar pattern, a sterile side line. The anoplotheres are often placed close to the anthracotheres, but with little positive reason. The long-legged xiphodonts are surely members of my tylopod assemblage but their closer relationships are matters of doubt. Simpson, like some earlier writers, bracketed them with the Camelidae, presumably because of their bodily proportions; Viret considers this as parallelism and would place them closer to the base of the selenodont stocks. The anoplotheres are placed by Simpson, followed by Viret, in a common superfamily with the anthracotheres, but there does not seem to me to be any strong evidence for this association. Colbert (636) has argued for a position of the Amphimerycidae ancestral to the traguloids, in which he is followed, with some doubt, by Simpson; Viret, on the other hand, would place the amphimerycids close to the xiphodonts on a lower evolutionary level.

The term Pecora is usually restricted to families of advanced selenodonts which blossomed out in the Miocene and Pliocene—cervids, giraffids, bovids, antilocaprids. The later Tertiary and Recent Tragulidae represent a morphologically antecedent stage. To the tragulids and some extinct groups which appear to be on the same evolutionary level the term Tragulina or Traguloidea is generally applied. There is no term in common use to include the seemingly natural group of pecorans proper plus their traguloid ancestors, and I have here used Pecora in a broader sense than usual. That this action should be taken is also suggested by the fact that one may doubt that the four "proper" pecoran families are a unit; the prongbucks may have originated independently from some primitive "traguloid" stock.

The Tragulidae themselves are not, of course, the direct ancestors of the higher pecorans and must be regarded as a side branch, since they do not appear in the fossil record until the development of more advanced families was already under way. The small group of Gelocidae of the late Eocene and Oligocene of Europe are considered by all to be traguloids, although probably not ancestral to "higher" families. Apart from this, the forms generally considered as constituting an ancestral pecoran group are those often included in the Hypertragulidae, with the curious Protoceratidae as a side branch. Colbert (particularly in 636) has carefully discussed pecoran ancestry and I have tended to follow him in this area (although I have not, as I probably should have, placed the very primitive genus *Archaeomeryx* in a separate family). Gazin, however, in his treatment of American Eocene artiodactyls has raised an interesting problem. He considers the Hypertragulidae, in a narrow sense, to have arisen directly from homacodont palaeodonts, but believes the *Leptomeryx* group (plus Protoceratidae) to have had a very distinct origin from a common ancestor with camels and oreodonts.

At about the beginning of the Miocene there burst forth a startling radiation of higher pecoran families—deer, giraffes, bovids, and prongbucks. There is no place among fossil vertebrates (save perhaps teleosts) where I feel less confident than here, and I have in general done no more in my text than list genera in the families to which they are usually assigned. There are, of course, problems about various primitive and aberrant forms that are unsettled. Particularly troublesome is the position of various early horned types: were the "horns" shed or bare, or if horn-covered, was the horn shed? Frequently a series of forms antecedent to deer (but possibly antecedent to other forms, particularly giraffes, as well) is separated as a family Palaeomerycidae. I have followed this precedent, although Simpson places most of these forms in primitive subfamilies of the Cervidae and a few in a family Lagomerycidae related to the giraffes. Viret follows Simpson in including most of the *Palaeomeryx*-like forms in the Cervidae, but also places *Lagomeryx* and its kin in the Cervidae, not in the giraffe group. Simpson erects three superfamilies—Cervoidea for the deer, Giraffoidea for the *Lagomeryx* group and giraffes, Bovoidea for the Bovidae and Antilocapridae. Since it seems rather generally agreed that the deer and giraffe lines are interwoven at their bases I have united them in the Cervoidea. I have retained the superfamily Bovoidea, although there is no positive proof that cattle and prongbucks, which have in common only the presence of horn sheaths and high-crowned teeth, are actually related; the two families may have evolved in parallel fashion in the two hemispheres, east and west.

Crusafont-Pairo describes the Giraffoidea in the Piveteau Traité (1961). Although (in contrast to Simpson) he does not include the *Lagomeryx* group, he does include here, as the discrete family Dromomerycidae, a series of American genera usually classed among the palaeomerycids or the deer proper and creates still another family for the poorly known European Miocene genus *Triceromeryx*. Among the bovids, I have done nothing except faithfully list the many scores of accepted fossil genera—this with the suspicion that a considerable reduction in numbers of genera could well be made.

Artiodactyl origins are less clear than those of the perissodactyls. As Schaeffer (619) has pointed out, the double-keeled artragalus is the key "invention" which is diagnostic of the order (and may be primarily responsible for artiodactyl success). Since our knowledge of the great majority of early Tertiary mammals is confined to dentitions and, at the most, skull materials, it may be a long time before we can trace the evolution of pre-artiodactyl types in which this structure was developing in Paleocene times. Schaeffer has pointed out that some members of the hyopsodontid condylarth family are dentally suggestive of artiodactyl relationship, and further evidence, skeletal as well as dental, may eventually prove that the artiodactyls are actually descended from the Hyopsodontidae.

25

Edentates

In early years the term "edentates" was used to cover pangolins and aardvarks as well as the numerous and varied South American forms. But the situation had been sufficiently clarified by the time of Simpson's 1945 work to allow the use of Edentata as a proper ordinal term for the Xenarthra, the South American edentates, plus their primitive North American relatives, the Palaeanodonta.

A considerable number of works on xenarthrans have appeared in the last two decades. Among South American workers, Castellanos, Paula Couto, and Cabrera have been active in this field. Resulting from a stay in Ecuador, Hoffstetter (651, etc.) turned his attention from the Squamata to this group, and in addition to descriptive papers has discussed general matters of edentate classification. Paula Couto (655) completed and published an unfinished work of Matthew describing the interesting pygmy ground sloths of Cuba.

In my classification of the Xenarthra there are several radical changes from that of Simpson. I have considered the armadillos and glyptodonts to form an infraorder Loricata (or Cingulata). I have added a separate family of armadillos for the recently discovered and aberrant *Pseudorophodon*; Hoffstetter (1954) would go farther and place *Pseudorophodon* in a discrete superfamily, distinct from both armadillos and glyptodonts.

I have added, following Hoffstetter (1954), a third superfamily, in addition to those for armadillos and glyptodonts, for *Palaeopeltis*. This form was regarded by Simpson as a glyptodont of dubious position. However, Kraglievich and Rivas (1951) (who use the generic term *Orophodon*) demonstrate that in *Palaeopeltis* we have an early Tertiary form that shows many features comparable to those of the sloths, but with a surface dermal armor. Much more data are needed regarding forms of this general category, but the presence of such a morphologically intermediate type, plus the fact that the oldest xenarthrans are armored, strongly suggest that the basal stock of the Xenarthra consisted of primitive, armored, armadillo-like forms, and that the "hairy" tree sloths, ground sloths, and ant-eaters have had a secondary loss of armor.

Anteaters, sloths, and ground sloths are very often regarded as making up a single infraorder Pilosa. However, the anteaters (poorly known as fossils) would appear to be quite remote from the sloths, and I have followed Hoffstetter (1954), in separating them as an independent infraorder Vermilingua.

It has been customary to cleave the remaining Pilosa, the sloths, into two units, one for the tree sloths, the other for the giant ground sloths. Patterson (438), however, points out that the Megalonychidae and Megatheriidae appear to be closely related, and that of the two living tree sloths, *Choloepus* approaches the megalonychids and *Bradypus* the megatheres. In contrast, mylodont sloths appear to be equally remote from the other ground sloths and both tree sloths. I have therefore considered the Mylodontidae to be the sole members of one superfamily of Pilosa, as the Mylodontoidea, with a second superfamily, Megalonychoidea, including the two other major ground-sloth families and both living tree sloths.

On the two related ground-sloth families, Megalonychidae and Megatheriidae, recent work suggests that many forms often assigned to the former actually pertain to the latter. When such a transfer is made, it would appear that the megalonychids differentiated at a late date from a common stock with the megatheres, and it is of interest that the Megalonychidae not only include the West Indian genera but also forms recorded from the United States as early as the early Pliocene and possibly late Miocene. This distribution suggests that a center of megalonychid evolution lay in the warmer parts of the northern continent (perhaps Mexico and Central America particularly?).

A curious recent find was that by Chow (1963) of an incomplete lower jaw, named *Chungchienia*, startlingly similar to that of a ground sloth, in the Eocene of China. Both in time and space we are far, far removed from any situation in which we would expect a ground sloth. Possibly this specimen may be a member of the peculiar Taeniodontia, also early Tertiary in age but otherwise confined to North America.

Aardvarks.—In recent years we have gained some additional knowledge of the geographical range of members of the Tubulidentata—in East Africa (MacInnes 569), Algeria (Arambourg 1959), the Caucasus (Burchak-Abramovich 1952). We have not, however, advanced our basic knowledge of the nature and relationships of the group since the publication of Colbert's (570) study of a Pontian skeleton from Samos. The origin of the group is still obscure. The hypothesis of condylarth relationships, strongly supported by Colbert, is still as good a surmise as any; certainly any relationships to the Edentata proper seems out of the question. Presumably here, as in certain other cases, the story will remain a blank until (if ever) we become better acquainted with the earlier Tertiary of Africa, where the aardvarks surely evolved. *Tubulodon* from the Lower Eocene of Wyoming was described by Jepsen in 1932 on the basis of fragmentary remains including teeth of spongy structure suggestive of those of *Orycteropus*; but Gazin (466) has pointed out that this form is probably a palaeanodont edentate.

Pangolins.—As in the case of the Tubulidentata, there is no reason to believe that the Pholidota of modern Asia and Africa have any relationship to the Edentata proper; these scale-covered creatures have nothing in common with the South American anteaters except for a parallel development of a long toothless snout associated with a diet of "ants." A few fragmentary remains from the European Tertiary have long been attributed (with doubt) to this order; no new data of importance have come to light in recent years, and the history and relationships of the order remain essentially blank.

26

Whales

In earlier decades of the century, much active work was carried on regarding fossil cetaceans, notably by Kellogg. Little has been accomplished further during the past two decades. Perhaps the possibility of new finds in the "classic" cetacean-bearing deposits has been essentially exhausted; presumably we must await the discovery and exploitation of new beds of appropriate nature before any great advances are possible. Meanwhile, there are major unsolved problems. There is a considerable body of information regarding each of the three suborders—Archaeoceti, Odontoceti, and Mysticeti—but the three groups are quite distinct, without connecting links. We know nothing of the ancestry of the two "modern" suborders, and, as Kellogg has pointed out, none of the known archaeocetes appear to lie in an ancestral position. Further, there is a complete blank as to the origin of the order as a whole. The oldest whales of the Eocene, although more primitive than later types, have nevertheless already assumed the major characteristics of the order. Until recently the oldest known forms were of middle Eocene age; the recent discovery of *Anglocetus* (Tarlo 1963) from the Lower Eocene of England pushes the story back a third of an epoch, but the remains of this form are too scanty to give us any new clues about origins. The fact that most of the earliest whale remains are from the shores of Africa suggests here, as in the case of the subungulates, that we are dealing with a group which developed on that continent. But from what antecedent forms—from carnivores of one type or another, or directly from some insectivore stock? Van Valen (543), on the basis of dentition, suggests comparison with the Hyaenodontidae or Mesonychidae, particularly the latter; but such comparisons cannot, of course, be considered too seriously in the absence of more positive evidence of transitional forms. Further, the question has been raised, by Kleinenberg (1959) and others, as to whether the Cetacea are actually a monophyletic group.

27

Rodents

Rodents proper.—Most abundant of mammals in terms of genera and species, both in the present fauna and in that of the Tertiary, and most perplexing of all groups in their evolution and classification are the rodents. The study of fossil rodents is a specialized one, from which many mammalian paleontologists tend to shy away. In recent times Schaub and Lavocat have been major European students of rodents; in North America, Gidley did much work on them early in the century, followed later by Burke and R. W. Wilson; currently A. E. Wood is a major rodent specialist, with Black and Dawson as more recent entrants into the field. Modern workers are much indebted to Ellerman's (661) monumental work for basic data on living rodents. During the past two decades a considerable number of valuable faunal studies have been published; we may cite, for example, Bohlin's (468) work on Kansu rodents, Dehm (1950) on the Burdigalian of Eichstätt; Wilson (478, 666) on early Tertiary rodents of North America and the Miocene of northeastern Colorado; various accounts by Kowalski of late Pliocene and Pleistocene faunas from Poland, and by Hibbard on faunas comparable in age from North America. Shotwell (667) has reviewed the aplodontids and mylagaulids, and Black (668) the North American sciurids. Wood's (664) monograph on paramyids is of major importance in the consideration of rodent evolution, as is that of Wood and Patterson (673) on the South American Oligocene fauna in a consideration of caviamorph evolution and classification. Stehlin and Schaub's (665) lengthy discussion of "trigonodonty" is responsible for establishment of one modern attempt at interpretation of rodent evolution. Lavocat (471, 663, etc.) has published valuable discussions on the position of Old World forms formerly included in the hystricomorphs.

Until recently the classification of rodents (and their presumed phylogeny) has been based primarily on Tullberg's study of the group in 1899 and even, in considerable measure, on that of Brandt in 1855. Using mainly the differentiations in skull architecture associated with the development of the powerful masseteric musculature (with an "assist" from the development of the angular region of the jaw), the rodents for many decades were grouped into three suborders, commonly termed Sciuromorpha, Hystricomorpha, and Myomorpha, with squirrel, porcupine, and rat as typical members. To push all rodents into these three categories was a matter of considerable difficulty (the bathyergids were perennially problematical) but up to two decades ago most writers valiantly strove to maintain this system (as did both Simpson and I in 1945).

There have, however, been dissenters from time to time, and the voice of dissent (and the arguments backing it) has constantly increased, so that today it is agreed that the former apparent simplicity in classification cannot be maintained.

Miller and Gidley in 1918 proposed a rather different system in which the major components were actually not too different from those of Brandt and Tullberg, but with a seemingly arbitrary lumping of miscellaneous types in a "superfamily" Dipodoidea. Schaub's classification, based upon studies by Stehlin and himself, as set forth in the Piveteau Traité (1958), is primarily based on cheek tooth structure. It is, in some regards, radically different from any other. Emphasis is placed on the presence or absence of a pentalophodont type of molar; rodents are divided into two suborders, Pentalophodonta and non-Pentalophodonta, the former including essentially the types usually placed in the Hystricomorpha plus the castoroids, with all other rodents classified as non-pentalophodonts (but this group essentially divided into two parts roughly corresponding to sciuromorphs and myomorphs).

An extreme example of a classification based on relationships, to which an excellent worker may be driven by pessimism, is that given by Thaler (1966) in which he divides rodents into ten suborders and even so leaves half-a-dozen families "*incertae sedis*." Not quite as extreme is a type of classification worked out in great measure by Wood, but with much of which Lavocat and various other workers seem to agree; this I have followed in my text. Wood has varied somewhat in his point of view in various papers (cf., for example, 662, 671, 672) but his essential thesis is that while a considerable proportion of rodents can be placed in one of several suborders, a number of families and even superfamilies cannot be closely associated with major groups and must, at least for the present, be left *incertae sedis*.

To a considerable degree his point of view is based on his thorough studies of Eocene rodents, particularly the Paramyidae (664), which he believes with obvious justice to represent the most primitive (and oldest) of known rodents. Seemingly quite surely descended from the paramyids, and still essentially primitive in many ways, are several other extinct families (such as the Sciuravidae, Cylindrodontidae, Mylagaulidae and the Aplodontidae). This assemblage, of which the sewellel is the only living survivor, Wood terms the Protrogomorpha (a term dating from Zittel [1893] but much modified as used by Wood). Although Wood currently does not include them in this suborder, the Sciuridae are obviously offshoots of this primitive stock, and I have departed mildly from Wood's current classification to include the squirrels (cf. Black 668), and thus make available the familiar term Sciuromorpha for the primitive rodent stock, with the squirrels as somewhat aberrant survivors.

Certainly the rat-mouse group, as in nearly every scheme of classification, is central in a second major subdivision, presumably derived along one line (perhaps several lines) from "protrogomorphs"; for them the term Myomorpha may be retained. There would, again, be majority agreement that a certain number of other rodent types might be allied with the Muridae and Cricetidae;[1] such types are the Dipodidae and Zapodidae, mainly Old World forms, and the American Geomyidae. (The last, with which the fossil Eomyidae appear allied and which may have come from a primitive stock through the Sciuravidae, were formerly considered to be sciuromorphs.)

[1] In my second edition I was persistently conservative in considering cricetids to be merely a murid subfamily; but apart from the fact that in general the two stocks seem distinct although closely allied, the vast numbers of forms included here make it advisable to separate the two at the family level.

With some doubt, the four Old World families of Gliridae (Myoxidae), Seleviniidae, Spalacidae, and Rhizomyidae may be appended to the Myomorpha.

A major change in current opinion has to do with the great assemblage generally termed the Hystricomorpha, characterized especially by the presence of a large canal enclosing part of the masseter running forward from the orbit into the face. Generally included here in older classifications were the Old World porcupines (*Hystrix*, etc.); a variety of odds and ends of other living Old World—African—rodents, such as the Anomaluridae, Pedetidae, Thryonomyidae, Ctenodactylidae, and Petromyidae; and some extinct Old World families, notably the Pseudosciuridae and Theridomyidae. The great bulk of the hystricomorph suborder, however, consisted of the numerous and varied forms which until the Pleistocene constituted the entire rodent population of South America (and of which the North American porcupine, *Erethizon*, is a recent northward wanderer).

Are we dealing, in this supposed suborder, with a natural group, divided geographically into South American and Old World components? The distribution argues strongly against it. Beliefs about a land bridge between Africa and South America in the Tertiary have today become unacceptable, and the South Atlantic is a long swim for a necessarily pregnant female. We have no knowledge of any North American form which could be a common ancestor of Old and New World components. It hence seems plausible to believe that the two sets of "hystricomorph" members have developed in parallel fashion, and that the supposed suborder Hystricomorpha is not a natural group.

Although to some extent Wood's (672, 673) thesis resembles that of the Queen of Hearts, "Sentence first, evidence afterward," his conclusion (1955) with which Lavocat (663) agrees, that the South American rodents should be treated as a separate suborder, Caviomorpha[2] (the guinea pig taking the place of the porcupine, with squirrel and rat, in the trio of major rodent "types"), is seemingly well justified. Wood and Patterson (673) have recently published a study of the oldest (Oligocene)

[2] Schaub suggested the term Nototrogomorpha.

South American rodents, with implications about the interrelationships of the dozen or so families making up the Caviomorpha. The oldest forms are Oligocene in age; their late appearance (together with the absence of representatives of the group in North America) give some slight (but very slight) support to those who would hold to the idea of transatlantic migration. But it seems probable that the earlier forms may have been in existence in the more tropical parts of South America in the earlier times, or may have been making a delayed journey by "island hopping" down through Central America in Eocene and Oligocene days.

We are now, however, left with a considerable series of odds and ends of rodents not assignable to any of the three major suborders. In older classifications, the beavers, Castoridae, were accepted as derived, like the squirrels, from ancestral sciuromorph types, and hence could be properly included in the same suborder. Presumably they are of paramyid derivation, but there are no connecting forms and they have diverged so far that Wood would place the castorids (together with the related *Eutypomys*) in a separate superfamily without subordinal assignment. Much more serious is the plight of the Old World forms formerly included in the hystricomorphs, as well as the highly specialized bathyergids (sometimes associated with them, although questionably). Lavocat, particularly, is a current student of fossil forms in this category and has published a number of valuable papers in this area (partly summarized in 663). He believes that the Pseudosciuridae and Theridomyidae, of the late Eocene and, mainly, the Oligocene of Europe, comprise (as a superfamily) a compact group of Old World forms of paramyid derivation (somewhat paralleling the caviomorphs); and I understand that further work by Lavocat, not yet published, suggests that *Phiomys* and abundant related types from the Oligocene and, especially, from the Miocene of Africa form the center of a second stock of onetime "hystricomorphs" of superfamily grade to which the cane rat, *Thryonomys*, and the rock rat, *Petromys*, belong. We are still left, however, with a number of other fragments, some occasionally attached questionably in former classifications to the Sciuromorpha, others to the erstwhile Hystricomorpha—the Old World porcupines, Hystricidae; the gundis, Ctenodactylidae; the African flying squirrels, the Anomaluridae; the springhare, *Pedetes*. In my ignorance, I fondly hope that eventually many of these fragmentary lists of families and superfamilies will be found to be related, with the eventual reestablishment of an Old World true hystricomorph group.

Lagomorphs.—Although the hares and rabbits were long classed as a suborder of rodents, their ordinal distinctness was realized by Gidley and other workers a half-century and more ago. So slow, however, has this realization spread among biologists other than paleontologists and mammalogists that one still finds today works on physiology or biochemistry (for example) in which studies made on "rodents" often prove to have been done on rabbits. The oldest known lagomorph, *Eurymylus* of the late Paleocene of Mongolia, is already a rather fully developed member of the order and gives no clues as to origins. Certainly there is nothing in the structure or dentition of *Eurymylus* to suggest even the most remote relationship to the Rodentia on an ordinal or even superordinal level. Van Valen (1964) has recently suggested that the lagomorphs are related to the deltatheridian creodonts. This is as good a guess as any other, but based almost entirely on the shape of the narrow v-shaped upper molars. Little attention has been paid to the group in recent years except by Dawson, who has done good work on its systematics (675, 1965, 1967), and by Wood (674), and by Tobien (1963).

28

Vertebrate History: Introductory, Paleozoic Vertebrates

Paleogeography.—I have discussed to some degree in my text the questions of ancient land bridges and possible continental positions in older times differing from those seen today. I was brought up in an atmosphere in which it was maintained that continents had always been in the exact position in which they are found today, and my early beliefs tended to be reinforced by Matthew's (435) demonstration that the numerous Tertiary land bridges proposed by various authors rested on very insecure foundations. Workers in mammalian paleontology have reinforced Matthew's point of view about the Tertiary and have tended to extrapolate backward in time, assuming that if the evidence refuted proposals for Tertiary land bridges or continental shifts, the same would hold true for earlier periods (with which they are less familiar).

With my own interests centered on older periods, however, I have found myself weakening in my early beliefs as to continental fixity and gradually falling away from the "true faith." In earlier years the freshwater fish faunas of the Devonian of Europe and North America appear to be in strong contrast with one another; today, with fuller knowledge, it appears, for example, that nearly three-quarters of the Devonian osteichthyan genera are likewise present in Europe—a much higher percentage than is true of the modern fauna. Again, it was earlier believed that the Carboniferous amphibian faunas of Europe and North America differed greatly; further study and comparison indicate a high degree of similarity and, as noted later, the early Permian tetrapod faunas of the two continents appear to be quite similar. Although I do not believe that the evidence is—as yet—strong enough to make a very positive statement, I can say that Paleozoic faunal distributions could be much more readily understood if the Atlantic Ocean had not been in existence in Paleozoic days, and Eastern and Western Hemispheres had been apposed.

For the northern continents, the evidence from the Mesozoic is not at all decisive, and the same is true of the Jurassic and Cretaceous periods as regards African—South American relationships. In the south, however, there is evidence strongly suggesting (although not proving) that in the Triassic there was free "trans-Atlantic" faunal interchange between Africa and South America. Gradually accumulating evidence tends to indicate faunal similarities between the two continents in early and middle Triassic days much greater than would be expected were only extremely circuitous northern connections between them present at that time. This, in turn, brings in the Gondwanaland concept, first brought forward by the paleobotanists, which assumes that in Paleozoic and earliest Mesozoic days peninsular India, Australia, and Antarctica, together with Africa and South America formed a single great southern continent. A thorough discussion of the Gondwanaland question took place in an international symposium held in Argentina in October, 1967.

Cambrian

No vertebrates of any sort are known in the Cambrian. Possibly the ancestors were then still in a protochordate stage, without skeletal parts capable of fossilization unless in such unusual circumstances as the Burgess Shale of Walcott or, if already in a vertebrate stage with hard parts developing, existing in an environment other than the marine beds that constitute the bulk of known Cambrian sedimentary deposits.

Many years ago Howell, in studying the Cambrian of northwestern Vermont, came upon a tiny plate which he was unable to identify as pertaining to any invertebrate group. On this negative evidence he concluded that it must be a vertebrate, and turned it over to Bryant (1927), who described it as "*Eoichthys howelli.*" However, Bather, an echinoderm expert, in a note (p. 40) in Smith Woodward's 1932 English revision of the Zittel volume dealing with lower vertebrates, identified the "dawn fish" as an echinoderm plate.

Ordovician

As far as I am aware, there are few if any definite reports of Ordovician sediments of continental nature, and hence no positive claim can be made for the presence of vertebrates in fresh water. Ordovician marine formations are widespread and frequently highly fossiliferous as regards invertebrates; vertebrate fossils, however, are extremely limited in distribution. A few toothlike structures, presumably dermal ornaments, were long ago discovered in the glauconitic sands of the Leningrad region and have been recently discussed by Ørvig (77). Apart from these, all known vertebrates are present in a line running south to north along, roughly, the Rocky Mountain front in Colorado and onward into Wyoming and Montana; Ørvig (77) has been a recent worker on materials from these areas. The older finds consisted mainly of innumerable tiny heterostracan fragments from the Harding Sandstone of Colorado; similar materials have been found at two horizons in the Bighorn Mountain region of Wyoming, and a find has been made in a well core in the Williston oil basin in Montana. These sediments have been thought to be nearshore deposits; a positive element is believed to have been present east of this area of deposition. The vertebrates may be interpreted as either (*a*) forms which lived in coastal marine waters or (*b*) forms which lived in continental fresh waters to the east (from which no sediments have been preserved) and whose fragmentary remains drifted out into the coastal region to the west.

Silurian

In earlier times the Downtonian of the Welsh border region of Great Britain and equivalent horizons elsewhere were included in the Silurian. In recent decades the general trend has been to consider the Downtonian and equivalents as Devonian, and in consequence the amount of known Silurian vertebrate data has been considerably reduced. Since the nature of these vertebrate-containing deposits has been a major bone of contention among those debating fresh- or saltwater origins of vertebrates, accounts of the Silurian vertebrate deposits are rather fully described in various papers concerned with this question, such as Romer and Grove (1935), Gross (1950), Romer (37),

Robertson (1957), White (1958*b*), and Denison (38). The vertebrates included, except for scales and spines of acanthodians, are all ostracoderms. A Bohemian find of placoderm material described by Gross (105) was originally thought to be late Silurian but, I understand, is now thought to lie above rather than below the Silurian-Devonian boundary.

Silurian sediments are abundant and well known in North America and Eurasia. Most are typically marine in nature; and, I believe, correlated with their nature is the fact that vertebrates are sparsely represented. In the United States much of the interior of the country contains numerous areas of marine Silurian exposures, in none of which vertebrates have ever been found. Toward the east, in the general Appalachian region, are beds which appear to represent nearshore and presumably deltaic deposits (Amsden 1955). Here, in contrast to the typical marine areas, are found heterostracan remains, mainly in "redbeds." A few remains of vertebrates are known from two areas in the maritime provinces of Canada; Denison (84) has recently reported Silurian pteraspids in marine sediments from Alaska; Thorsteinsson (1958) has found on Cornwallis Island in the Arctic abundant heterostracan fossils, now in process of description, which appear to be in marine sediments, and Dineley (1966) has also found Arctic Silurian vertebrates on Somerset Island.

In Europe, again, marine Silurian sediments are abundant but, again, vertebrate remains are limited as to areas and horizons. Notable as regards vertebrates are (*a*) the English-Welsh border; (*b*) southern Scotland; (*c*) the Ringerlike region of Norway; and (*d*) the island of Oesel in the Baltic. There are numerous excellent reports on the Welsh borderlands—several by White (1950, 1961) are important as regards vertebrates, and Allen and Tarlo (1963) have recently published a stratigraphic summary. The classic study of the Scottish area is by Peach and Horne (1899), but this is in need of revision, and fishes are present in redbeds far below the Downtonian—of Wenlockian or even earlier age (Ritchie, in Rolfe 1961). For Norway, see Kiaer (70) and Heintz (64); the Oesel succession is described by Hoppe (1931). The English, Scottish, and Norwegian beds show, as one ascends the series, a gradual shift, with fluctuations, from marine conditions to estuarine and deltaic sediments, and, finally, to continental redbeds, with an increase in vertebrate life as we progress toward the end of the succession. In the English-Welsh area the Upper Silurian Ludlow group, apparently a nearshore deposit, shows, occasionally, fragmentary remains of acanthodians and coelolepids in the form of scales and spines (Squirrel 1958). At the summit of the Ludlow group is the Ludlow bonebed, with abundant if fragmentary vertebrates; this is now generally held to mark the base of the Devonian and the approach of continental sedimentation of "Old Red" type. In Scotland and Norway a gradual and fluctuating shift toward continental conditions takes place somewhat earlier, so that articulated vertebrate materials, notably anaspids, are found in the late Silurian. The Oesel sequence is different in nature; vertebrates are present in a sequence denoted by the letter K. K^1 at the base appears to represent lagoonal deposits with a fauna essentially fresh water in nature, notable for osteostracans such as *Thyestes*. K^2 and K^3 are marine, and vertebrates are represented only by isolated scales and fragments. K^4 is a complex, multilayered series; it is mainly marine, and vertebrates are mainly present in occasional layers which are essentially a series of intercalated bonebeds.

No Silurian vertebrates are known in the southern continents.

To summarize: certain Silurian finds are definitely in marine rocks, but many of these consist of fragmentary materials only and many represent, it seems, nearshore deposits into which animals from fresh waters may have drifted. The best case for marine habitat is that of the Heterostraci. In general, however, we may say of Silurian vertebrates that *finds increase as we approach continental conditions geographically or stratigraphically*.

Seemingly highly significant is the fact that of the numerous major types of gnathostomes which were to appear early in the Devonian—varied placoderms, crossopterygians, lungfishes, actinopterygians—not one trace has been found to date in the Silurian. The only jawed fishes present are acanthodians, and these are represented by fragmentary materials alone. The absence of any trace of other jawed fishes in the Silurian can be explained only (1) by the assumption that the ancestral forms were without skeletal hard parts capable of fossilization (White 1958*b*) and that all types, essentially simultaneously, blossomed out into a bony condition, or (2) by the more reasonable assumption that the preserved Silurian record, as known, does not include sediments laid down in the environment in which these ancestral forms lived. For those who, like myself, believe the early center of vertebrate life to have been in fresh waters, the explanation is simple—the dearth of freshwater sediments in the Silurian. For those who advocate salt water, the problem is rather more difficult, but may be explained (as by Denison 38) by the assumption that the ancestral forms lived in some specialized and restricted environment, such as nearshore marine areas.

Devonian

A great amount of valuable work has been done on Devonian vertebrates during the past two decades, but relatively few new fish-bearing beds of this age have been discovered during this period. Summaries of Devonian fish occurrences are given in papers cited for the Silurian dealing with early vertebrate environments. Known Old World Devonian fish-bearing deposits were long confined almost entirely to Europe; recently, however, late Devonian arthrodires have been described from Morocco (Lehman 99), finds have been made in the Lower Devonian of northern and eastern Asia by Russian and Chinese workers, and further Devonian material has been described from Australia (White 93). In North America a considerable fauna has been discovered in the Utah Lower Devonian (Denison 68, 79, 92). The black shales of the Cleveland, Ohio, region have long been famous for abundant materials of arthrodires and cladoselachians; currently, extensive excavations for new highways are yielding large amounts of excellent new specimens.

Carboniferous

I have followed current American custom in treating the long stretch of Carboniferous time as comprising two distinct periods, Mississippian and Pennsylvanian. The two are quite distinct in nature, as is the case with the nearly equivalent Lower and Upper divisions of the Carboniferous in Europe. The Mississippian, or Lower Carboniferous, is dominantly marine; the Pennsylvanian and the European Upper Carboniferous are, of course, notable for the presence of coal deposits. The early history of tetrapod life is inadequately known because of the paucity of continental deposits of Mississippian age. In Europe, amphibian remains of this age are present only in Scotland. The fauna is a meagre one, but certain of the lepospondylous types have very recently been restudied by the Broughs of Cardiff (1967). I have recently redescribed the only known Lower Carboniferous labyrinthodont skeleton, that of *Pholidogaster* (Romer 201), and Panchen (1964) has redescribed the important *Palaeogyrinus* skull. A limited amount of Mississippian material is now known from West Virginia, mostly undescribed, but otherwise tetrapods of this age are almost unknown in North America, and none at all are known from the remaining continents. The dearth of materials here is the most serious gap in our current knowledge of continental tetrapod history (except perhaps for the Jurassic).

Much better, but far from adequate, is our knowledge of Pennsylvanian tetrapods. In Europe a moderate amount of material has long been known from Great Britain, much of which was summarized by Watson (199, 200). The "Gaskohle" fauna of Bohemia was early described by Fritsch (172); much of it was reinterpreted by Steen (213), but reexamination in the light of modern knowledge is highly desirable. In North America major early finds were from the tree stumps of the Joggins cliffs of Nova Scotia, from the cannel of Linton, Ohio, and from the nodules of the Mazon Creek region of Illinois. Moodie in 1916 attempted a résumé of these materials, but his essay was extremely faulty. Both Steen (1931) and I (1930) restudied the Linton material; I corrected various mistakes of Moodie, but made a number on my own behalf, so that further study is here advisable. The Joggins material is being restudied with care by Carroll (239, 1966, 1967*a*) with valuable results. In 1955 I initiated a new series of attempts to discover new North American Pennsylvanian tetrapod materials. This, continued by Baird, has had results of value in Nova Scotia; new specimens from this area are now in process of study and description.

Fish faunas from the Carboniferous are abundant. Perplexing, however, are the numerous finds of teeth and spines of "sharks" from marine beds, notably from Mississippian limestones of both Great Britain and the Mississippi Valley. It is possible that at least a fraction of the problems set us by these fragmentary remains will be solved by the fish remains being studied by Zangerl from black Carboniferous shales from Indiana (cf. Zangerl and Richardson 1963).

Permian

The position of the Carboniferous-Permian boundary is a question which has greatly perturbed invertebrate paleontologists, who generally assume that the fixing of period-system boundaries lies in their domain. Their concern has been essentially in determining the base of the Permian. This system, based on Russian beds, was established originally in but a vague fashion and did not at first include beds which covered anywhere near the entire extent of the Permian as we now conceive of this system; further, there has been considerable debate among Russian workers about where the base of the system should be placed. In consequence, American workers, at least, have tended to abandon the Russian sequence, establish a "neotype" for the Permian in the marine beds of trans-Pecos, Texas, and decree that the base of the system lies at the base of the Wolfcamp group in that region.

Actually, however, determination of this boundary (important for vertebrate workers) does not properly lie within the sphere of activity of the invertebrate paleontologists, but in that of the paleobotanists. It is, properly, the upper boundary of the Carboniferous, not the base of the originally vague Permian that is to be sought. That point, at which the continental coal measures are considered to give way to the basal Permian Rothliegende, can be determined only on botanical evidence. This matter long underwent thorough consideration by European stratigraphers and paleobotanists, and in a Heerlen conference in 1933 (Jongmans and Gothan 1937) it was decided to draw the line below the point at which *Callipteris conferta* appears in the flora. American invertebrate workers seem to have been in general quite unaware of this; however, their west Texas Permian base is probably not far from the established European summit of the Carboniferous. In north central

Texas the late Pennsylvanian-Permian includes both marine and continental elements. The equivalent of the base of the Wolfcamp is thought to lie somewhat below the summit of the Harpersville Formation (Cheney 1940; King 1942). Extensive work on the floras of this region was done by C. B. Reed. This has not been published, but a note in the paper by King cited above (p. 648) states that the Newcastle coals at a fairly low level in the Harpersville have a flora of late Pennsylvanian type and that *Callipteris* (species not stated) is present in the upper part of this formation.

For the early Permian the most productive continental beds are those of northern Texas, where the formations include the Wichita and Clear Fork groups and have, over the past 90 years, yielded a wealth of reptiles, amphibians, and fishes. I published a summary of our knowledge of them in 1958 (in Westoll 14, pp. 157–79). To the north, these beds continue through Oklahoma into southern Kansas. In recent years, Olson (383, etc.) has extended our knowledge of fossils from this region upward from the Clear Fork into the lower part of the Double Mountain (Pease River) group, and is extending our limited knowledge of the Oklahoma beds by current exploration (Olson 1965, etc.).

A second American region of increasing importance lies farther west in the Abo and Cutler formations, found from New Mexico and Colorado into eastern Utah and northern Arizona. Fossils have been long known from these beds in New Mexico and have recently been discovered farther north and west, in areas currently being explored by Vaughn (Lewis and Vaughn 1965; Vaughn 1964, etc.). Minor American continental deposits are those of the Dunkard of the upper Ohio Valley (Romer 1952) and of Prince Edward Island (Langston 1963).

Closely comparable to the American early Permian beds are the redbeds of the Rothliegende of Germany and similar beds in France and England. The material is much smaller in amount and variety than in America, but it seems highly probable that the two faunas were basically similar (Romer 1945); this leads one to doubt the existence of an Atlantic barrier in Permian times.

Until recent decades one neglected early Russian Permian vertebrate finds and "jumped" to South Africa, where the thick but barren Ecca beds of the early Permian are followed by the Beaufort Series, the Permian members including the abundant fossiliferous Tapinocephalus, Endothiodon, and Cistecephalus zones of the Karroo. The evolutionary stage of the fossils of even the Tapinocephalus Zone is clearly more advanced than that of the typical Texas fauna; a definite gap existed.

This gap is now being closed. More than a century ago a few Permian vertebrates were described from Russia; further Permian finds (not well described at the time) were made by Amalitzky in the 1890's; in the last three decades a very considerable amount of Permian material has been found and described. Much of the data has been summarized in English by Olson (383). Four Permian zones were distinguished. The highest, Zone IV, is equated with the Cistecephalus Zone of South Africa; Zone III is problematical; Zone II is essentially comparable to the Tapinocephalus Zone of the Karroo; Zone I is somewhat earlier and has rather more primitive forms. This last essentially closes the gap between the early Permian beds of America and those of South Africa, for Olson (383) finds in the highest fossiliferous American beds (of the Double Mountain or Pease River) forms comparable to those of the Russian Zone I.

In earlier years there was a general trend to divide the Permian in tripartite fashion; as regards continental beds, those of America and western Europe were regarded as early Permian, the Tapinocephalus Zone and Russian Zones I and II as middle Permian, the Endothiodon and Cistecephalus zones of the Karroo and Russian Zone IV as later Permian. A current trend is toward a merely bipartite division. This is followed by Olson who, however, would cleave the period at an extremely early point, considering that not only all the Russian and South African fossiliferous zones but even the Double Mountain of America are Upper Permian. I would strongly disagree with this conclusion. It makes a very uneven division when one considers the total picture of Permian vertebrate evolution.

29

Vertebrate History: Mesozoic Vertebrates

Triassic

Our knowledge of the sequence of Triassic continental vertebrate faunas has, until recently, been greatly handicapped because of the dearth of appropriate continental sediments in the Northern Hemisphere. The period name is, of course, derived from the three successive types of sediments found in central Europe—Bunter, Muschelkalk, Keuper. The first of the three consists essentially of solidified mud banks with little tetrapod fauna except such strays as have floated into the near-shore deposits of the southern Alps, worked principally by Peyer and in recent years by Kuhn-Schnyder. Even in the late Triassic continental redbeds of the Keuper, little is known except for amphibians until the relatively late Stubensandstein is reached, with a good fauna of saurischian dinosaurs. (Schmidt's compilation of 1928 and his 1938 supplement give a detailed account of the central European Triassic faunas.) In North America there was until recent decades little trace of tetrapods other than amphibians until the Upper Triassic redbeds, such as the Newark, Chinle, and Dockum, were reached. The upper part of the Newark, particularly in the Connecticut Valley, shows a wealth of footprints proving the presence of a dinosaur fauna comparable to that of the Keuper; the western formations have yielded few dinosaurs, but abound in remains of metoposaurian labyrinthodonts and phytosaurs, also present in the European Keuper.

For early Triassic tetrapods, we long had no adequate representation except in South Africa. Here the upper part of the Beaufort beds, proved equivalent to the Scythian marine zone of the basal Triassic through the nature of the contained amphibians and fishes, is present as the Lystrosaurus and Cynognathus zones (the so-called Procolophon Zone is an ecological variant of the Lystrosaurus Zone, *fide* Hotton and Kitching

1963). Here the fauna is composed almost entirely of therapsids —cynodonts, dicynodonts, and bauriamorphs—together with a minor proportion of other forms, notably several members of the Thecodontia, the ancestral archosaurs from which the dinosaurs were to evolve. Following the Beaufort in South Africa come the Molteno beds, presumably Middle Triassic, but long thought to be barren of vertebrates, and above this the redbeds and Cave Sandstones of the late Triassic, with a dinosaur fauna comparable to that of the Keuper. Work in China has revealed early and late faunas comparable to those of the Beaufort and Keuper, respectively, but lacking, as in other cases, a Middle Triassic assemblage.

Until the last few decades, thus, our knowledge of Triassic faunas was almost entirely confined to an early assemblage, dominantly therapsid in nature, and a late one, dominantly dinosaurian. A middle component was essentially missing. All that could be inferred concerning its probable nature was that in middle Triassic times the therapsid fauna would presumably have been declining and the thecodonts advancing to approach or reach the dinosaurian level anticipatory of the Keuper.

In the last quarter-century our horizon has widened considerably. To gain a knowledge of the middle portion of the period we must turn to Africa and South America. In the former continent, the Manda beds of East Africa have yielded a considerable fauna, studied by Huene and by Parrington and his students. At long last the Middle Triassic Molteno beds have yielded a limited number of specimens to Crompton. Much more is now known of the Middle Triassic in South America. The Santa Maria beds of southern Brazil have an abundant fauna collected by Huene (238) and further collections have been made by L. I. Price for the Brazilian Geological Survey and by Harvard and the American Museum of Natural History. Proving to be still better are beds in western Argentina, particularly in the Ischigualasto-Talampaya region. Here is a series of formations, in part fossiliferous, extending from the Permian to the late Triassic. Below redbeds (Los Colorados Formation) from which Bonaparte is describing dinosaurs and other reptiles of Keuper type is the Ischigualasto Formation, yielding an abundant fauna of Middle Triassic type (Romer 1962*b*; and faunal papers by Bonaparte and other authors).

Putting together the data from the Manda, Molteno, Santa Maria, and Ischigualasto formations, we now have an adequate picture of the vertebrate assemblage of the Middle Triassic—at least as regards the Southern Hemisphere. As expected, the fauna shows a decline in the percentage of therapsids present and an advance in thecodont evolution toward the dinosaur level, reaching that level in the form of a relatively small amount of material of true dinosaurs. But in addition, these beds show two positive characteristics that could not have been foretold. (1) In the Lower Triassic, a few of the cynodonts present appear to have departed from the primitive carnivorous mode of life, to acquire broadened chewing "gomphodont" molar dentitions. In the Middle Triassic, normal cynodonts are still present, but in reduced numbers; the gomphodonts, on the other hand, are immensely abundant. (2) The earliest rhynchocephalians of the Lower Triassic include two forms which were tending to produce multicusped toothplates in the upper jaw and a slicing blade below. These forms, as rhynchosaurs, had become highly abundant in the Middle Triassic; together with the gomphodonts they make up perhaps three-quarters of all specimens collected at Ischigualasto, for example.

As a result of this work, we have a picture of the Triassic faunal succession which, simplified (or, rather, oversimplified), is as follows.

A. Lower Triassic. Dominantly therapsid in nature; early thecodonts and minor elements of other sorts.
B. Middle Triassic. Dominantly gomphodonts and rhynchosaurs; thecodonts advancing toward and to the dinosaur level.
C. Upper Triassic. Dominantly dinosaurian, with advanced and specialized thecodonts.

The complete picture is, of course, more complex; there are intergradations between the elements of this faunal succession and regional differentiations. Metoposaurs and phytosaurs are present in the northern continents, although Baird has pointed out to me that they appear not to have survived until the end of the Triassic; there are, for example, no traces of their presence in the late Newark footprint deposits. The Maleri beds of India are considered to be Upper Triassic in age; but in addition to phytosaurs and metoposaurs, rhynchosaurs are present, suggesting a relatively early Upper Triassic level. The Newark beds may cover a considerable time interval; beds presumably of early Newark age in Nova Scotia contain rhynchosaurs (Baird and Take 1959), and Baird informs me that a gomphodont is present as well. Rhynchosaurs have long been known in Great Britain in beds correlated with the Keuper. But in that region the Muschelkalk is absent; the so-called Keuper redbeds presumably are in part Middle Triassic, and part, at least, of the rhynchosaur finds may well be of that epoch. The Triassic of Elgin, Scotland, contains rhynchosaurs. The Elgin fauna is currently being restudied by Walker (283, 322), who correlates it with the continental Upper Keuper because of the presence of closely related aetosaurid thecodonts. No rhynchosaur remains have ever been discovered in the continental European Keuper, however; the one dinosaur known from Elgin, *Ornithosuchus*, is so primitive that it has previously been assigned to the Thecodontia. The Elgin fauna may represent a "B-C" intermediate.

In South America the recently discovered Chañares fauna, two formations below the Ischigualasto, shows an abundance of rather primitive gomphodonts but apparently no rhynchosaurs; it is presumably an "A-B" transition (Romer 1966, 1967*a*). From southern Mendoza Province, Bonaparte is describing a gomphodont and forms which he cannot distinguish from *Cynognathus* and *Kannemeyeria* of South Africa; we have here, obviously, an "A" fauna. The Cacheuta series of Mendoza Province, mainly studied by Rusconi (cf. Romer 1960) is deficient in reptilian remains but may also be in part at the "A" level.

Jurassic

A splendid record of marine vertebrates is present in the Jurassic, notably in the early Jurassic Lias of England, Normandy, and Bavaria, and in the late Jurassic lithographic limestones of Germany and France. Continental faunas, however, are poorly represented except in the late Jurassic Tendaguru beds of Tanzania and, especially, the widespread late Jurassic Morrison beds of North America. For the greater part of the Jurassic we know little of land forms except such strays as have floated out to be deposited in lagoonal or estuarine saltwater deposits. One of the greatest desiderata (apart from the early Paleozoic) in the vertebrate story would be a full-fledged terrestrial fauna from the earlier Jurassic—a time when, it is obvious, important advances were taking place in the evolution of certain of the reptilian groups and, most especially, notable events were surely occurring in mammalian evolution. In most regions our chances of filling this gap seem none too good. In South America and Africa the great cycle of deposition of continental beds which had begun in the Permian and continued into the Triassic came to an end in late Triassic times. Little future development can be hoped for, it would seem, in North America or in Europe, despite the fact that fissure deposits in western England and southern Wales have recently yielded a number of interesting finds of Rhaetic-Liassic age. Asia, however, is still relatively unexplored, and in India Pamela Robinson, with the cooperation of Indian workers, is currently exploring continental beds—the Khota Formation—of Liassic age, where it is hoped that a representative fauna will be found (Jain, Robinson, and Chowdhury 1962).

While our main concern in the Jurassic is for fuller knowledge of continental tetrapods, there is as well a major lacuna in our knowledge of fish evolution in that period. We have a wealth of Jurassic fishes but nearly all are saltwater forms and the story is thus an unbalanced one. What events were occurring in fresh waters? One tends, for example, to visualize the origins and early radiation of teleosts to have taken place in the sea. But were fresh waters of Jurassic times relatively destitute of higher actinopterygians? We can account, not unreasonably, for many Cenozoic fish inhabitants of fresh waters as being reimmigrants from the seas. But what, for example, of the great carp-catfish group, the Ostariophysi, which are in Tertiary and Recent times almost exclusively freshwater forms? Are they ultimately of marine origin? Or, rather, may they not have arisen in the Mesozoic in the fresh waters in which they now live? We know almost nothing of their early history, and it is probable that better Jurassic (and Cretaceous) continental finds will give the answer.

Cretaceous

In many regards the Cretaceous story of aquatic vertebrates is rather fully known. There are abundant marine beds with a host of fishes and marine reptiles. The situation is rather unbalanced, however, for it is the late Cretaceous that is best represented, notably by the widespread chalk deposits, and our knowledge of both marine fishes and aquatic reptiles is much poorer for the early Cretaceous than for the later parts of the period. Further, for the Cretaceous, as for the Jurassic, data on freshwater fishes are woefully inadequate.

A similar lack of balance is present in the array of tetrapods from continental deposits. Famous are the Upper Cretaceous dinosaur beds of western North America and of Mongolia; the latter, first made known through American Museum expeditions in the 1920's, have been reexplored by Russian and, more recently, Polish expeditions (Efremov 1954*a*, 1956*b*; Kielan-Jaworowska and Kowalski 1965). The continental beds of the early Cretaceous, however, are relatively poorly known. The Wealden beds of the English Channel region have, of course, long been famous, but in other regions beds of early Cretaceous age are either poorly developed or inadequately explored. In North America the discovery of mammals in the Trinity sands has called attention to these Texas deposits, which deserve further work. Although perhaps little advanced over the Morrison in the nature of their faunas, the early Cretaceous Cloverly beds of Wyoming are potentially of great interest; vertebrates were collected there many years ago by Barnum Brown, and current collections being made there by Ostrom should yield valuable results.

But even in the productive beds of the late Cretaceous, finds have mainly been of large forms, notably, of course, dinosaurs, and our knowledge of small reptiles and amphibians and of mammals of the Cretaceous has lagged behind. In early years Hatcher's technique of sifting anthills yielded a certain amount of material, and in recent years careful sifting of materials on a considerable scale has begun to yield valuable results. Patterson's work on the Trinity sands (Patterson 502; see Slaughter 1965) has increased greatly our former minute knowledge of early Cretaceous mammals; extensive sorting of Lance materials is greatly increasing our knowledge of Lance lower vertebrates as well as mammals (McKenna 1962, 1965*a*; Clemens, 452, 1966; Estes, 269), while the fabulous "Bug Creek" locality in Wyoming (Sloan and Van Valen 451) is yielding exceedingly important data on early placental history.

30

Vertebrate History: Cenozoic Vertebrates

Tertiary continental connections.—As noted earlier, Matthew, Simpson, and other workers have successfully demolished all, or nearly all, supposed Cenozoic "land bridges" and sunken continental connections suggested by a variety of workers in the last century and in the earlier decades of the present one. Quite surely nearly every faunal puzzle can be explained by the make-or-break of such intercontinental connections as currently exist or may have existed in the not-distant geological past; many cases of discontinuous tropical distribution can be explained by former existence of the groups concerned in the connected Holarctic region; most supposed bridges constructed for a single group prove absurd when the faunal picture is considered as a whole.

But while the interrelations of the continents may have been much as today during most or all of the Cenozoic, it cannot be completely taken for granted that continental positions have remained the same throughout Cenozoic time. If one entertains the possibility of there having been continental drift in Paleozoic or even Mesozoic times, might it not be that such drifting had not been completed by the end of the Mesozoic? Even ardent supporters of drift would generally agree that Wegener was wrong in his assumption that the development of the North Atlantic was a relatively recent affair. But attention should be called to Kurtén's (1966) recent discussion of the early Eocene faunas of North America and Europe. At this time the faunas of the two continents were exceedingly similar, much more so than at any later period of the Tertiary, suggesting to Kurtén that there was some closer connection between these areas than that supplied by the roundabout route via Asia. Kurtén believes that some connection existed across the northern Atlantic area. An advocate of continental drift might suggest that in the earliest Tertiary the "rift" between eastern and western hemispheres had not been completed, and that a contact still persisted at the north between the two.

Another suggestion as to a somewhat different continental configuration in the Tertiary has been brought forward by Australian workers. It is generally assumed that Australia has always (or at least for a long time) had its present position close to the East Indian chain. However, under the hypothesis that there once existed a major southern continent, a super-Gondwanaland, its fragmentation would have produced not only South American and African portions, and a persistently southern Antarctica, but fragments traveling east and north to become peninsular India and Australia. Under this hypothesis, Australia may well have been sailing northeastward during much of the Tertiary, to have "grounded" only relatively recently on the southern margin of the East Indian region. Such a hypothesis offers a number of interesting suggestions about Australian faunal history. It may be, for example, that the Australian marsupial fauna, marooned in the Indian Ocean aboard this great floating vessel, was even more isolated in its development than we generally imagine. It may have been only recently (geologically) that the gap between the Indies and Australia became narrow enough for limited faunal exchange to have occurred. The interesting rodent fauna of Australia (Simpson 1961) is often assumed to have reached that continent by "island hopping" far back in the Tertiary. But, alternatively, under the hypothesis just discussed, this invasion may have been relatively late, followed by rapid differentiation.

Recent advances.—The past two decades have seen much active work in the exploration of Cenozoic beds. A sample of numerous important faunal papers published during recent years is included in references numbered 442–94 in my *Vertebrate Paleontology*. In Europe and North America advances have been mainly in broadening our knowledge of beds and faunas already known, such as, in North America, Jepsen's accumulation of considerable amounts of material, mostly unpublished as yet, on the late Paleocene and early Eocene of the Bighorn Basin, Gazin's fieldwork and studies in other Wyoming areas, Simpson's (1948) work on the Eocene San José Formation of the San Juan Basin, much work on the Pacific Coast Tertiary by the Berkeley and, more recently, Los Angeles groups. Hibbard (488, 492) has, by the screening method, greatly broadened our knowledge of late Pliocene and Pleistocene faunas. Shotwell has in progress a very broad-range study of the late Tertiary of a region in eastern Oregon.

Notable in Europe is the energetic and productive work of Crusafont-Pairo and his collaborators in exploring the Spanish Tertiary. Russian workers have found much Tertiary material in western "Siberia" and they (and more recently Polish expeditions) are adding to the knowledge of the early Tertiary as well as the Cretaceous of Mongolia. Highly noteworthy is a great amount of exploration of Cenozoic beds in China by the Peking laboratory, with finds described mainly by M. M. Chow. In northern Africa, Arambourg's work on Ternifine (1963) is a major contribution to the late history of mammalian life in North Africa. Work on the Fayûm beds, long in abeyance, has been energetically resumed by Simons. Leakey's continued work in East Africa is opening up a considerable field of possibilities in that region, notably at Olduvai (Leakey 494) and at Rusinga (on which the British Museum has already published a number of papers, such as nos. 517, 536, 569). It would appear that a considerable variety of horizons are present in East Africa which may eventually give us a broad picture of "Neogene" history in Africa. In South America, notable recent discoveries include a prolific late Paleocene fissure site explored by Paula Couto (460), and the Miocene La Venta deposits in Colombia explored by Stirton and his colleagues (Stirton 1953). Simpson (459) in 1948 published part of a comprehensive study of the early Tertiary mammal faunas of South America; publication of a second, concluding part of the study is expected in 1968. Pleistocene faunas in a La Brea type of preservation have been found and described in the oil-bearing coastal regions of Peru and Ecuador by Hoffstetter (1952) and by Edmund and Churcher (Churcher 1959; Edmund 1965). As noted earlier, Stirton (510) is responsible for a renewal of interest in attempts to resolve the problem of Tertiary history in Australia.

Epochs and ages.—By many European workers the Cenozoic is considered to consist of two periods, Paleogene or Nummulitic and Neogene, the former corresponding to the Paleocene-to-Oligocene of general English and American usage, the latter to the Miocene and later epochs. More general, however, is the use of epoch names—Paleocene, Eocene, etc. Establishment of such a series of epochs defined by European marine faunas took place well back in the last century. Mammalian workers (as well as workers on marine invertebrates) have, in both Europe and North America, attempted to subdivide these epochs into a series of ages (or stages), and in most cases a sequence is well established. There are, however, many difficulties in equating continental and marine faunas. In certain instances in Europe, notably in France and the Rhineland, continental-marine equivalencies are well established, but even in that continent numerous problems remain. In North America most Tertiary continental beds are far inland, and only on the Gulf Coast and California is it possible, to a limited degree in the later Cenozoic, to equate continental and marine beds. As a consequence of the paucity of interdigitations, it is probable that in many instances continental stages are not exactly equivalent to marine stages of approximately the same age, and, most especially, equivalencies between European and North American stages have been established by direct comparison of continental mammalian faunas rather than by attempts to work through marine equivalencies. Stages roughly parallel to those for Europe and North America have been established for Asia and South America; in general these have been erected purely on the basis of the succession of continental mammalian faunas, without reference to marine faunas. Thaler (1965) has essayed the construction of a similar series, based on continental mammalian forms from Europe.

In 1941 a committee under the chairmanship of H. E. Wood published a valuable report (446) on the nomenclature and correlation of the North American Tertiary, and established a series of well-defined stages which have won general acceptance. In 1959 preparation of a new report was begun under the same chairman; owing, however, to the illness of Wood, the report, taken over by Simpson and Patterson, has not yet been published. Despite the presence of various points of disagreement, it is probable that few changes in the stage sequence will be made.

Paleocene.—The Paleocene was a late addition to the series of Cenozoic epochs; there was little in the European marine sequence to merit the erection of a separate pre-Eocene epoch, and many European geologists ignore the term, lumping such equivalents in with the Lower Eocene. The North American continental sequence shows clearly that a considerable time period is involved, however, and a distinct Paleocene series and epoch is warranted (although it is quite probable that both Cretaceous-Paleocene and Paleocene-Eocene boundaries as currently established in the North American continental sequence are higher than in the accepted marine sequence). Three major subdivisions are clearly indicated—Puercan, Torrejonian, and Tiffanian—to which the 1941 committee report added a Dragonian age for a single fauna, intermediate between Puercan and Torrejonian, from the North Horn Formation of Utah, and a final Clarkforkian for a fauna close to that of the Wasatchian. A recent study by R. C. Wood (1966), however, demonstrates that the Clark Fork fauna is essentially Wasatchian and that this stage name should be eliminated. A recent thorough study of the European Paleocene faunas has been published by Russell (461). The well-known Cernay fauna is clearly late Paleocene. That from Walbeck in Germany is somewhat earlier in appearance, and Russell suggests that it be referred to the Middle Paleocene; there appears, however, little evidence that it is markedly older than that of Cernay.

Miocene.—Considerable differences of opinion have long persisted, and still persist, in Europe over the boundaries of the Miocene. The marine Aquitanian appears faunally to be late Oligocene; on the other hand, the continental faunas referred to the Aquitanian (and possibly not exactly equivalent) may well be early Miocene. I have so tabulated them and equated them (together with the earlier Burdigalian) with the Harrison and other beds of the Arikareean stage. At the top of the Miocene sequence, the Pontian (or Pannonian) "*Hipparion*" fauna, widespread from western Europe through Greece and Samos to Persia, is regarded by some European workers as late Miocene, by others as early Pliocene. From an American point of view, it would appear to be essentially comparable to the Clarendonian and hence Lower Pliocene.

The interesting faunas found by Leakey at Rusinga Island, etc., are generally assumed to be Lower Miocene in age. This has been questioned by Evernden *et al.* (1964); as Simpson (1967) points out the situation is complex, and much further data need to be gained before we can work out the sequence in the interesting—and potentially highly important—series of Tertiary beds of Kenya.

Pleistocene.—The limits of the Pleistocene have been, and probably will continue to be, matters for dispute. The time of its termination should obviously be fixed at some point in the final retreat of the last glaciation, and we are probably approaching a condition where, thanks to Carbon 14, in addition to other evidence, we can decide upon a date in terms of actual calendar years. More difficult has been the fixing of a Pliocene-Pleistocene boundary. As a general proposition, the Pleistocene should include the time covered by the glaciations of Europe and North America—the Gunz-Mindel-Riss-Würm of the Alps and the presumably equivalent sequence of eastern North America. But despite the evidence gained from study of river terraces and beach levels, it is in many cases difficult to establish correlations with glaciations, even in the non-glaciated areas of the northern continents, more difficult in the tropics, where the main geologic evidence is confined to the determination of two major pluvials and an interpluvial, and still more difficult again in the southern continental regions.

A major change in recent years has been the proposal, discussed and accepted by the Eighteenth International Geological Congress in London in 1948 (Oakley 1950) to extend the lower limit of the Pleistocene downward to cover a considerable period preceding the onset of glaciation, to include the Villafranchian-Calabrian of Europe. This very greatly expands the extent of the Pleistocene. It includes the entire time since the appearance of *Equus*, which is thus available as a marker for the Pleistocene; on the other hand, it introduces into the early part of the Pleistocene, as expanded, a number of forms formerly regarded as purely Tertiary (such as *Hipparion* and certain primitive mastodonts). Kurtén (1963) has recently summarized the nature of the Villafranchian mammal fauna.

In North America a Blancan was established in the 1941 committee report as a final stage of the Pliocene, essentially equivalent to the Villafranchian. If the Villafranchian is to be included in the Pleistocene, so, logically, should the Blancan. This, however, has met with strong opposition from some major workers on Pleistocene mammals, such as Hibbard, who believe that at least part of the Blancan faunas—such as the type Blanco and the well-explored Rexroad of Kansas—should be retained in the Pliocene.

In South America, exact correlations are difficult to make. It seems probable that the Chapadmalal is essentially a Villafranchian–Blancan equivalent, but current South American workers prefer to retain the Chapadmalalian in the Pliocene.

The Pleistocene, of course, marks in many ways the climax of the history of mammals. There is in most groups a general trend toward increasing size as the Cenozoic proceeds. The Pleistocene is notable as marking a size climax. Apart from the presence of numerous large mammals now extinct, one gains the impression that even forms which have survived were represented by individuals larger in the Pleistocene than those now living; in some cases this has been definitely demonstrated by metrical studies such as those of Hooijer and Kurtén. In addition, we find in the Pleistocene extinct forms of giant size related to modern types but generically distinct from the types which have survived—giant beavers both in Europe and North America, giant armadillos in the Americas, giant echidnas in Australia, and so forth.

The disappearance during the Pleistocene of varied types of large mammals of stocks now extinct is, of course, well known. This is apparent to some extent in Eurasia (as in the case of the extinction of mammoths, woolly rhinoceroses) and in Africa, in which the Pleistocene included survivors of a number of archaic groups. The extinction phenomena in the Americas are more striking. In South America we find in the Pleistocene representatives of the native ungulate stocks still present, although in reduced variety, and glyptodonts and ground sloths, destined for extinction. In both Americas there were native equids and proboscideans. Most startling is the contrast between Pleistocene and Recent faunas in North America. Today the list of large mammals in this continent can be practically told off on one's fingers; in the Pleistocene there were, in addition, a score or more of genera of large mammals now extinct.

The time of extinction is a matter of interest. One might, a priori, assume that in the northern continents, at least, climatic vicissitudes might have been responsible for a gradual extinction during the Pleistocene. But there is evidence suggesting late survival for part of the doomed forms. In Europe, for example, cave paintings demonstrate the presence of the woolly mammoth and woolly rhinoceros in relatively recent Upper Paleolithic times, and there is evidence of survival of certain South American extinct forms until the (relatively late) arrival of man.

Of especial interest is the time factor in North America. In the early decades of the twentieth century, the major authority on Pleistocene mammals was the late Oliver P. Hay. He had found a good representation of the large types, later extinct, in an early Pleistocene (supposedly Aftonian) deposit, whereas there was no evidence satisfactory to him of the presence of such forms as camels and horses in later deposits in the glaciated areas. Accordingly, he claimed that the main extinction in North America took place in the early Pleistocene. This created a major problem as regards archaeological finds and a long-standing feud between Hay and his contemporary, Hrdlicka, a self-appointed watchdog on presumed early human finds. Time after time, reports would be made of artifacts in association with elements of the extinct Pleistocene fauna. If the association were correct, this would mean the presence of *Homo sapiens* in North America in early Pleistocene, Aftonian, days. This was obviously absurd, and hence Hrdlicka tended to deny the reality of the supposed association.

My attention was called to the matter by there being brought to me the skull of a native camel from a Utah cave—a skull so fresh that there were still bits of dried meat on it. Obviously it had not been preserved in this fashion since the Aftonian—at least half-a-million years ago—and I undertook a general review of our knowledge of the time element in relation to North American Pleistocene mammals (Romer 485). I strongly suspected that Hay's basic thesis was incorrect, and that part, at least, of the ancient fauna had survived until later. The most reasonable hypothesis was that a gradual extinction had occurred during the extent of the Pleistocene. Surprisingly, the data when summed up took still another form. Although our knowledge of extinct faunas was then much less than it is today, and although such aids as Carbon 14 were not then available, the evidence seemed clearly to indicate that practically the entire fauna of large mammals had persisted through all the vicissitudes of the Pleistocene and only became extinct in essentially Recent times—a conclusion strongly confirmed by all later studies.

Why did this great extinction occur during or at the close of the Pleistocene? And why was this mainly an extinction of large forms? In view of the late survival of many elements, at least, the glacial conditions can have played but a small part in the story. With modern interest in radiation, a large "dose" of cosmic rays has been suggested (as in the case of the dinosaurs) as a cause. But again, as in the case of the dinosaurs, this does not account for the selective elimination of large beasts. In South America, the influx from North America of carnivores and of better-adapted herbivores among ungulates and rodents may have been influential. But (man apart) faunal irruptions can have played but little part in other continents. In North America, for example, no new element of importance entered during the Pleistocene except the bison; this may have reduced the importance of antilocaprids but probably had no further effect.

Can man have played an important role in reduction in, for example, North America? Man arrived here at a late Pleistocene date, in time for the faunal extinction. It is possible that his intrusion may have in some fashion upset a delicate ecological balance. I have suggested that his presence may have been influential by disturbing breeding habits, but this suggestion met with little favor. I did not think it probable that man played a major direct role by hunting and killing the great beasts; otherwise we would find many more examples of man-mammal association than we do. Martin (1966), however, argues for this point of view, pointing out that in Africa a considerable, if smaller, extinction of mammals took place in mid-Pleistocene times. This coincided with the time of attainment by man of a certain stage in development when, he thinks, he could and did strongly influence directly the mammalian fauna and was responsible for the African extinction. He would assume that the arrival of man in North America, at a similarly advanced stage even if in small numbers, was similarly responsible for the major extinction in this continent. Leakey (1966), however, promptly countered by pointing out that the African extinction coincided with the mid-Pleistocene interpluvial period in Africa, that it was climatic conditions—drought—which caused extinction there, and that Martin's analogy is not meaningful.

Bibliography

Numbered references in the text are to the Bibliography in my Vertebrate Paleontology, *3d. ed.* (*Chicago: University of Chicago Press, 1966*).

Allen, J. R. L. and Tarlo, L. B. 1963. The Downtonian and Dittonian facies of the Welsh Borderland. *Geol. Mag.* 100:129–55.

Amsden, T. W. 1955. Lithofacies map of Lower Silurian deposits in central and eastern United States and Canada. *Bull. Amer. Assoc. Petrol. Geol.* 39:60–74.

Appleby, R. M. 1961. On the cranial morphology of ichthyosaurs. *Proc. Zool. Soc. London* 137:333–70.

Arambourg, C. 1959. Vertébrés continentaux du Miocène supérieur de l'Afrique du Nord. *Publ. Serv. Carte Geol. Algér., Paléont. Mém.* No. 4. Pp. 161.

———. 1963. Le gisement de Ternifine. Part 2. L'*Atlanthropus mauritanicus*. *Arch. Inst. Paléont. Humaine, Mém.* 32:37–190.

Attridge, J., Ball, H. W., Charig, A. J., and Cox, C. B. 1964. The British Museum (Natural History)—University of London joint palaeontological expedition to Northern Rhodesia and Tanganyika, 1963. *Nature* (*London*) 201:445–49.

Auffenberg, W. 1959. *Anomalophis bolcensis* (Massalongo), a new genus of fossil snake from the Italian Eocene. *Breviora, Mus. Comp. Zool.* No. 114:1–16.

———. 1961. A new genus of fossil salamander from North America. *Amer. Midl. Natur.* 66: 456–65.

Baily, W. H. 1884. Some additional notes on *Anthracosaurus edgei* (Baily sp.), a large sauro-batrachian from the Lower Coal Meassures, Jarrow Colliery, near Castlecomer, County Kilkenny. *Rept. Brit. Assoc. Adv. Sci.* 53:496–97.

Baird, D. 1957. Rhachitomous vertebrae in the loxommid amphibian *Megalocephalus*. *Bull. Geol. Soc. Amer.* 68: 1698.

———. 1964*a*. *Changisaurus* reinterpreted as a Jurassic turtle. *Jour. Paleont.* 38:126–27.

———. 1964*b*. The aïstopod amphibians surveyed. *Breviora, Mus. Comp. Zool.* No. 206:1–17.

Baird, D. and Take, W. F. 1959. Triassic reptiles from Nova Scotia. *Bull. Geol. Soc. Amer.* 70:1565–66.

Beer, G. de 1954. *Archaeopteryx lithographica.* Brit. Mus. (Nat. Hist.). Pp. 68.

Bendix-Almgreen, S. E. 1966. New investigations on *Helicoprion* from the Phosphoria Formation of south-east Idaho, U.S.A. *Biol. Skr. K. Danske Videnskab. Selsk.* 14(5):1–54.

Berg, D. E. 1966. Die Krokodile, insbesondere *Asiatosuchus* und aff. *Sebecus?*, aus dem Eozän von Messel bei Darmstadt/Hessen. *Abhandl. Hess. Landesam. Bodenforsch.* 52:1–105.

Bergounioux, F.-M. 1938. *Archaeochelys pougeti*, nov. gen., nov. sp., tortue fossile du Permien de l'Aveyron. *Bull. Soc. Géol. France* (5)8:67–75.

———. 1955. Testudinata. Chelonia. In: *Traité de Paléontologie* (ed. J. Piveteau), 5:487–544. Paris.

Bock, W. J. 1963. The cranial evidence for ratite affinities. *Proc. XIII Internat. Ornith. Congr.* (*1962*) 1:39–54.

Bohlin, B. 1953. Fossil reptiles from Mongolia and Kansu. *Rept. Sci. Exped. NW Prov. China* Publ. 37 (6):1–105.

Bonaparte, J. F. 1966. Un nueva "fauna" Triásica de Argentina. *Ameghiniana* 4:243–96.

Boonstra, L. D. 1962. The dentition of the titanosuchian dinocephalians. *Ann. S. Afr. Mus.* 46:57–112.

———. 1963. Diversity within the South African Dinocephalia. *S. Afr. Jour. Sci.* 59:196–206.

Brink, A. S. 1951. Studies of Karroo reptiles. I. Some small cynodonts. *S. Afr. Jour. Sci.* 47: 338–47.

———. 1960. A new type of primitive cynodont. *Palaeont. Afr.* 7:119–54.

———. 1963. The taxonomic position of the Synapsida. *S. Afr. Jour. Sci.* 59:153–59.

Brodkorb, P. 1963. Birds from the Upper Cretaceous of Wyoming. *Proc. XIII Internat. Ornith. Congr.* (1962) 1:55–70.

Brough, M. C. and Brough, J. 1967. Studies on early tetrapods. *Phil. Trans. Roy. Soc. London* (*B*) 252:107–65.

Bryant, W. L. 1927. Evidence of the presence of chordates in the Cambrian. *15th Bien. Rept. Vermont State Geolog., 1925–26.* Pp. 125–26.

Burchak-Abramovich, N. I. 1952. [On the fossil *Orycteropus gaudryi* from the Upper Tertiary deposits of the town Maraga.] *Vest. Gos. Mus. Gruznii* 15*A*:225–31 (In Russian.)

Butler, P. M. and Hopwood, A. T. 1957. Insectivora and Chiroptera from the Miocene rocks of Kenya Colony. *Brit. Mus.* (*Nat. Hist.*), *Fossil Mammals of Africa* No. 13:1–35.

Bystrow, A. 1956. [*Kolymaspis sibirica* g.n., s.n.—a new representative of Lower Devonian Agnatha.] *Vest. Leningrad. Univ. 1956* No. 18:5–13. (In Russian.)

Cabrera, A. 1947. Un saurópodo nuevo del Jurásico de Patagonia. *Notas Mus. La Plata, 12, Paleont.* No. 95:1–17.

Camp, C. L. 1923. Classification of the lizards. *Bull. Amer. Mus. Nat. Hist.* 48:289–481.

———. 1954. The Shoshone Mountain ichthyosaur site. Fallon, Nev. Pp. 2.

Carlsson, A. 1909. Die Macroscelididae und ihre Beziehungen zu den übrigen Insectivoren. *Zool. Jahrb. Abt. Syst.* 28:349–400.

———. 1922. Über die Tupaiidae und ihre Beziehungen zu den Insectivora und den Prosimiae. *Acta Zool.* 3:227–70.

Carroll, R. L. 1966. Microsaurs from the Westphalian B of Joggins, Nova Scotia. *Proc. Linn. Soc. London* 177: 63–97.

———. 1967*a*. Labyrinthodonts from the Joggins Formation. *Jour. Paleont.* 41:111–42.

———. 1967*b*. An adelogyrinid lepospondyl amphibian from the Upper Carboniferous. *Can. Jour. Zool.* 45:1–16.

Casamiquela, R. M. 1961. Dos nuevos estagonolepoideos Argentinos (de Ischigualasto, San Juan). *Rev. Asoc. Geol. Argentina* 16:143–203.

———. 1964. Sobre un dinosaurio hadrosáurido de la Argentina. *Ameghiniana* 3:285–312.

———. 1966. Nuevo material de "*Vieraella herbstii*" Reig. *Rev. Mus. La Plata* n.s., 4:265–317.

Casier, E. 1961. Matériaux pour la faune ichthyologique Eocrétacique du Congo. *Ann. Mus. Congo Belge, in-8°, Sci. Geol.* 39:1–93.

Chang, K.-J. 1965. New antiarchs from the Middle Devonian of Yunnan. *Vert. Palasiatica* 9:1–14.

Cheney, M. G. 1940. Geology of north-central Texas. *Bull. Amer. Assoc. Petrol. Geol.* 24:65–118.

Chow, M. M. 1953. Remarks on the two Mesozoic mammals from the northeastern provinces. *Acta Palaeont. Sinica* 1:150–56.

———. 1963. A xenarthran-like mammal from the Eocene of Honan. *Scientia Sinica* 12:1889–93.

Chowdhury, T. R. 1965. A new metoposaurid amphibian from the Upper Triassic Maleri Formation of Central India. *Phil. Trans. Roy. Soc. London* (*B*)250:1–52.

Chudinov, P. K. 1955. [Cotylosaurs from the Shikov-Chirkov site.] *Doklady Akad. Nauk URSS* 103:913–16. (In Russian.)

———. 1957. [Cotylosaurs from the Upper Permian Redbeds of the Urals.] *Trudy Paleont. Inst. Akad. Nauk SSSR* 68:19–87. (In Russian.)

Chudinov, P. K. 1960. [Upper Permian therapsids of the Yeshova locality.] *Paleont. Zhurn.* No. 4:81–94. (In Russian.)

———. 1964. Varanoidea, Cholophidia. *In: Osnovy Paleontologii* (ed. I. A. Orlov), 5 [Amphibia, Reptiles, Birds]:473–83. Moscow. (In Russian.)

Churcher, C. S. 1959. Fossil *Canis* from the tar pits of La Brea, Peru. *Science* 130:564–65.

Clemens, W. A., Jr. 1966. Fossil mammals of the type Lance Formation in Wyoming. Part II. Marsupialia. *Univ. Cal. Publ. Geol. Sci.* 62:1–122.

Cocude-Michel, M. 1960. Les sauriens des calcaires lithographiques de Bavière, d'âge Portlandien inférieur. *Bull. Soc. Géol. France* ser. 7, 2:707–10.

———. 1963. Les rhynchocéphales et les sauriens des calcaires lithographiques (Jurassique supérieur) d'Europe occidentale. Thèse Univ. Nancy, Fac. Sci. Pp. 187.

Colbert, E. H. 1948. A hadrosaurian dinosaur from New Jersey. *Proc. Acad. Nat. Sci. Philadelphia* 100:23–37.

———. 1951. Environment and adaptations of certain dinosaurs. *Biol. Rev.* 26:265–84.

———. 1961. The Triassic reptile *Poposaurus*. *Fieldiana: Geology* 14:59–78.

———. 1963. Fossils of the Connecticut Valley. The age of dinosaurs begins. *Bull. State Geol. Nat. Hist. Surv. Connecticut* No. 96:1–31.

———. 1964. The Triassic dinosaur genera *Podokesaurus* and *Coelophysis*. *Amer. Mus. Novit.* No. 2168:1–12.

———. 1965. A phytosaur from North Bergen, New Jersey. *Amer. Mus. Novit.* No. 2230:1–25.

———. 1966. A gliding reptile from the Triassic of New Jersey. *Amer. Mus. Novit.* No. 2246:1–23.

Colbert, E. H., Cowles, R. B., and Bogert, C. M. 1946. Temperature tolerances in the American alligator and their bearing on the habits, evolution and extinction of the dinosaurs. *Bull. Amer. Mus. Nat. Hist.* 86: 327–74.

Coon, C. 1962. *The Origin of Races*. New York. Pp. 724.

Cope, E. D. 1880. The skull of *Empedocles*. *Amer. Nat.* 14:304.

Crompton, A. W. 1964. A preliminary description of a new mammal from the Upper Triassic of South Africa. *Proc. Zool. Soc. London* 142: 441–52.

Crompton, A. W. and Ellenberger, F. 1957. On a new cynodont from the Molteno beds and the origin of the tritylodontids. *Ann. S. Afr. Mus.* 44:1–14.

Crompton, A. W. and Jenkins, F. A., Jr. 1967. American Jurassic symmetrodonts and Rhaetic "pantotheres." *Science* 155:1006–9.

Crusafont-Pairo, M. 1961. Giraffoidea. *In: Traité de Paléontologie,* (ed. J. Piveteau), 6(1):1022–37. Paris.

———. 1962. Constitución de una nueva clase (Ambulatilia) para los llamados "Reptiles mamiferoides." *Notas Comun. Inst. Geol. Min. Espana* No. 66: 259–65.

Danil'chenko, P. G. 1960. [Bony fishes of the Maikovski beds of the Caucasus.] *Trudy Paleont. Inst. SSSR* 78:1–208.

Dawson, M. R. 1965. *Oreolagus* and other Lagomorpha (Mammalia) from the Miocene of Colorado, Wyoming, and Oregon. *Univ. Colo. Studies Ser. Earth Sci.* No. 1:1–36.

———. 1967. Lagomorph history and the stratigraphic record. *In: Essays in Paleontology and Stratigraphy, Raymond C. Moore Commemorative Volume.* Univ. Kansas, Dept. Geol. Spec. Publ. 2:227–316.

Dehm, R. 1950. Die Nagetiere aus dem Mittel-Miocän (Burdigalium) von Wintershof-West bei Eichstätt in Bayern. *Neues Jahrb. Min. Geol. Palëont. Abhandl.* 91*B*:321–427.

Denison, R. H. 1951. The exoskeleton of early Osteostraci. *Fieldiana: Geology* 11:197–218.

D'Erasmo, G. 1946. L'ittiofauna Cretacea dei dintorni di Comeno nel Carso triestino. *Atti Accad. Sci. Napoli* ser. 3a, 2:1–136.

Dineley, D. 1966. Geological studies in Somerset Island, University of Ottawa expedition, 1965. *Arctic* 19(3):270–76.

Dobzhansky, T. 1963. Genetic entities in hominid evolution. *In: Classification and Human Evolution* (ed. S. L. Washburn). Chicago. Pp. 347–62.

Dollo, L. 1883. Troisième note sure les dinosauriens de Bernissart. *Bull. Mus. Hist. Nat. Belgique* 2:85–126.

———. 1923. Le centenaire des iguanodons (1822–1922). *Phil. Trans. Roy. Soc. London* (*B*)212:67–68.

Dorr, J. A. 1958. *Prouintatherium*, new uintathere genus, earliest Eocene Hoback Formation, Wyoming, and the phylogeny of Dinocerata. *Jour. Paleont.* 32:506–16.

Dowling, H. G. 1959. Classification of the Serpentes: a critical review. *Copeia* Pp. 38–52.

Dughi, R. and Sirugue, F. 1957. Les oeufs de dinosauriens du bassin d'Aix-en-Provence. *C.R. Acad. Sci. Paris* 245: 707–10, 907–9; 246: 2271–74, 2386–88 (1958).

Eaton, T. H., Jr. 1960. A new armored dinosaur from the Cretaceous of Kansas. *Univ. Kansas Paleont. Contrib., Vertebrata* No. 8:1–24.

Eaton, T. H., Jr. and Stewart, P. L. 1960. A new order of fishlike Amphibia from the Pennsylvanian of Kansas. *Univ. Kansas Publ., Mus. Nat. Hist.* 12:217–40.

Edinger, T. 1964. Midbrain exposure and overlap in mammals. *Amer. Zool.* 4:5–19.

Edmund, A. G. 1965. A late Pleistocene fauna from the Santa Elena Peninsula, Ecuador. Life Sciences Contribution, Royal Ontario Museum, Univ. Toronto, No. 63:1–19.

Efremov, J. A. 1939. First representative of Siberian early Tetrapoda. *Doklady Akad. Nauk URSS* n.s., 23:106–10.

———. 1946. On the subclass Batrachosauria—an intermediary group between amphibians and reptiles. *Bull. Acad. Sci. URSS, Cl. Math. Nat.* (*Ser. Biol.*) 1946:615–38. (Russian, English summary.)

———. 1954*a*. [Paleontological research in the Mongolian People's Republic. Preliminary results of the expeditions of 1946, 1948, and 1949.] *Trudy Mongol'skoi Komissii, Akad. Nauk USSR* 59:1–32. (In Russian.)

———. 1954*b* [Terrestrial vertebrate fauna from the Cupriferous Sandstones of the Permian of the Pre-Urals.] *Trudy Inst. Paleozool. Akad. Nauk URSS* 54:1–416. (In Russian.)

———. 1956*a*. [American elements in the fauna of Permian reptiles of the USSR.] *Doklady Akad. Nauk URSS* 111: 1091–94. (In Russian.)

———. 1956*b*. [Road of the winds. Notes from the Gobi Desert.] Moscow. Pp. 360. (In Russian.)

Estes, R. 1967. Anatomy and relationships of the primitive fossil snake *Dinilysia*. *Yearbook Amer. Phil. Soc.* 1966:334–36.

Evans, F. G. 1942. The osteology and relationships of the elephant shrews (Macroscelididae). *Bull. Amer. Mus. Nat. Hist.* 80:85–125.

Evernden, J. F., Savage, D. E., Curtis, G. H., and James, G. T. 1964. Potassium-Argon dates and the Cenozoic mammalian chronology of North America. *Amer. Jour. Sci.* 262: 145–98.

Fahlbusch, K. 1964. Die Stellung der Conodontida im biologischen System. *Palaeontographica* 123*A*:137–201.

Flerov, K. K. 1952. [Pantodonts collected by the Mongolian Paleontological Expedition of the Academy of Sciences, USSR.] *Trudy Inst. Paleont. Akad. Nauk URSS* 41:43–50. (In Russian.)

———. 1957. [Dinoceratia of Mongolia.] *Trudy Inst. Paleont. Akad. Nauk URSS* 67:1–82. (In Russian.)

Fox, R. C. 1962. Two new pelycosaurs from the Lower Permian of Oklahoma. *Univ. Kansas Publ., Mus. Nat. Hist.* 12:297–307.

Friant, M. 1952. Les chauves-souris frugivores firent-elles partie des mammifères les plus anciens? *Rev. Stomatol.* 53:207–11.

———. 1960. Les Haramyidae, mammifères du Rhétien d'Europe. *Ann. Soc. Géol. Nord* 79:212–21.

Gadow, H. F. 1933. *The Evolution of the Vertebral Column*. Cambridge, England. Pp. 356.

Gardiner, B. G. 1967. Further notes on palaeoniscoid fishes with a classification of the Chondrostei. *Bull. Brit. Mus.* (*Nat. Hist.*) *Geol.* 14:143–206.

Garstang, W. 1929. The morphology of the Tunicata and its bearing on the phylogeny of the Chordata. *Quart. Jour. Micr. Sci.* n.s., 72:51–187.

Gazin, C. L. 1965. A study of the early Tertiary condylarthran mammal *Meniscotherium. Smithson. Misc. Coll.* 149(2):1–98.

Gill, E. D. 1954. Ecology and distribution of the extinct giant marsupial "*Thylacoleo.*" *Victorian Nat.* 71:18–35.

———. 1965. The paleogeography of Australia in relation to the migrations of marsupials and men. *Trans. New York Acad. Sci.* ser. 2, 28:5–14.

Gilmore, C. W. 1946*a*. A new carnivorous dinosaur from the Lance Formation of Montana. *Smithson. Misc. Coll.* 106(13):1–19.

———. 1946*b*. Reptilian fauna of the North Horn Formation of central Utah. *U.S. Geol. Surv. Prof. Pap.* 210:29–53.

Gilmore, C. W. and Stewart, D. R. 1945. A new sauropod dinosaur from the Upper Cretaceous of Missouri. *Jour. Paleont.* 19:23–29, 540.

Ginsburg, L. 1961. La faune des carnivores Miocènes de Sansan (Gers). *Mém. Mus. Natl. Hist. Nat.* (*Paris*) new ser. C, 9:1–190.

———. 1967. Sur les affinités des mésosaures et l'origine des reptiles euryapsides. *C. R. Acad. Sci. Paris* 264:244–46.

Glikman, L. S. 1958. [On the classification of sharks.] Sep. Publ. A. P. Karpinsky Geol. Mus. Leningrad. Pp. 20. (In Russian.)

———. 1962. [Elasmobranch evolution, transgressive and regressive epochs.] *Vsesoyuznoe Paleont. Obshch.* 5–6:226–34.

Goin, C. J. and Goin, O. B. 1962. *Introduction to Herpetology*. San Francisco. Pp. 341.

Gregory, J. T. 1948. A new limbless vertebrate from the Pennsylvanian of Mazon Creek, Illinois. *Amer. Jour. Sci.* 246:636–63.

———. 1950. Tetrapods of the Pennsylvanian nodules from Mazon Creek, Illinois. *Amer. Jour. Sci.* 248:833–73.

———. 1952. The jaws of the Cretaceous toothed birds, *Ichthyornis* and *Hesperornis*. *Condor* 54:73–88.

Gregory, W. K. 1910. The orders of mammals. *Bull. Amer. Mus. Nat. Hist.* 27:1–524.

———. 1946. Pareiasaurs versus placodonts as near ancestors to the turtles. *Bull. Amer. Mus. Nat. Hist.* 86:275–326.

———. 1947. The monotremes and the palimpsest theory. *Bull. Amer. Mus. Nat. Hist.* 88:1–52.

Griffiths, J. 1956. Status of *Protobatrachus massinoti*. *Nature* (*London*) 177: 342–43.

———. 1963. The phylogeny of the Salientia. *Biol. Rev.* 38:241–92.

Gromova, V. 1960. [A new family (Tschelkariidae) of primitive carnivores (Creodonta) from the Oligocene of Asia.] *Trudy Paleont. Inst. Akad. Nauk SSSR* 77:41–74. (In Russian.)

Gross, W. 1937. Das Kopfskelett von *Cladodus wildungensis* Jaekel. 1. Teil: Endocranium und Palatoquadratum. *Senckenbergiana* 19:80–107.

———. 1947. Die Agnathen und Acanthodier des obersilurischen Beyrichienkalks. *Palaeontographica* 96*A*:89–161.

———. 1950. Die paläontologische und stratigraphische Bedeutung der Wirbeltierfaunen des Old Reds und der marinen altpaläozoischen Schichten. *Abh. Deutsch. Akad. Wiss. Berlin, Math.-Nat. Kl.*, 1949 No. 1:1–130.

———. 1953. Devonische Palaeonisciden-Reste in Mittel- und Osteuropa. *Paläont. Zeitschr.* 27:85–112.

———. 1954. Zur Conodonten-Frage. *Senckenbergiana Lethaea* 35:73–85.

———. 1957. Über die Basis der Conodonten. *Paläont. Zeitschr.* 31:78–91.

———. 1960. *Tityosteus* n. gen., ein Riesenarthrodire aus dem rheinischen Unterdevon. *Paläont. Zeitschr.* 34:263–74.

———. 1961. Aufbau des Panzers obersilurischer Heterostraci und Osteostraci Nord-deutschlands (Geschiebe) und Oesels. *Acta Zool.* 42:73–150.

———. 1962. Puet-on homologuer les os des arthrodires et des teléostomes? *In:* Problèmes Actuels de Paléontologie (Évolution des vertébrés). *Coll. Internat. Centre Nat. Recher. Sci.* No. 104:69–74.

———. 1963. *Drepanaspis gemuendenensis* Schlüter Neuuntersuchung. *Palaeontographica*, Abt. A 121:133–55.

———. 1963. *Drepanaspis gemuendenensis* Schlüter Neuuntersuchung *Palaeontographica*, Abt. A 121:133–55.

———. 1965*a*. Über die Placodermen-Gattungen *Asterolepis* und *Tiaraspis* aus dem Devon Belgiens und einen fraglichen *Tiaraspis*-Rest aus dem Devon Spitzbergens. *Bull. Inst. Roy. Sci. Nat. Belgique* 41:1–19.

———. 1965*b*. Über einen neuen Schädelrest von *Stensiöella heintzi* und Schuppen von *Machaeracanthus* sp. indet. aus dem Hunsrückschiefer. *Notizbl. Hess. Landesamt. Bodenforsch.* 93:7–18.

———. 1966. Kleine Schuppenkunde. *Neues Jahrb. Geol. Paläont. Abh.* 125:29–48.

Gurr, P. R. 1963. A new fish fauna from the Woolwich Bottom Bed (Sparnacian) of Herne Bay, Kent. *Proc. Geol. Assoc. London* 73:419–47.

Harris, J. E. 1951. *Diademodus hydei*, a new fossil shark from the Cleveland shale. *Proc. Zool. Soc. London* 120: 683–97.

Hecht, M. 1959. Amphibians and reptiles. *In:* The geology and paleontology of the Elk Mountain and Tabernacle Butte area,

Wyoming (ed. P. O. McGrew). *Bull. Amer. Mus. Nat. Hist.* 117:130–46.

Heintz, A. 1938. Notes on Arthrodira. *Norsk Geol. Tidsskr.* 18:1–27.

———. 1962. New investigation on the structure of *Arctolepis* from the Devonian of Spitzbergen. *Norsk Polarinst. Arb.* 1961 (1962):23–40.

Heller, F. 1960. Der dritte *Archaeopteryx*-Fund aus den Solnhofer Plattenkalken des oberen Malm Frankens. *Jour. Ornith.* 101:7–28.

Heyler, D. 1957. Révision des *Branchiosaurus* de la région d'Autun. *Ann. Paléontol.* 43:45–111.

———. 1958. Remarques sur la caudale d'*Acanthodes*. *C. R. Acad. Sci. Paris* 247:1636–39.

———. 1962. Les acanthodiens et le problème de l'Aphetohyoïdie. *In:* Problèmes Actuels de Paléontologie (Évolution des Vertébrés). *Coll. Internat. Centre Nat. Recher. Sci.* No. 104:39–47.

Hill, W. C. O. 1955. *Primates, Comparative Anatomy and Taxonomy. 2. Haplorhini: Tarsioidea.* Edinburgh. Pp. 347.

Hörstadius, S. 1951. *The Neural Crest.* London. Pp. 111.

Hofer, H. 1955. Neuere Untersuchungen zur Kopfmorphologie der Vögel. *Proc. XI Internat. Ornith. Congr.* Pp. 104–37.

Hoffstetter, R. 1952. Les mammifères Pléistocènes de la République de l'Équateur. *Mém. Soc. Géol. France* n.s. 31, No. 66. Pp. 391.

———. 1953. Les sauriens anté-crétacés. *Bull. Mus. Natl. Hist. Nat. Paris* ser. 2, 25:345–52.

———. 1954. Les gravigrades cuirasses du Déséadien de Patagonie (Note préliminaire). *Mammalia* 18:159–69.

———. 1955*a*. Squamates de type moderne. *In*: *Traité de Paléontologie* (ed. J. Piveteau), 5:606–62. Paris.

———. 1955*b*. Rhynchocephalia. *Ibid.* 5:556–76.

———. 1955*c*. Thecodontia. *Ibid.* 5:665–94.

———. 1964. Les Sauria du Jurassique supérieur et spécialemente les Gekkota de Bavière et de Mandchourie. *Senckenbergiana Biol.* 45:281–324.

Holmgren, N. 1933. On the origin of the tetrapod limb. *Acta Zool.* 14:185–295.

Hoppe, K. H. 1931. Die Coelolepiden und Acanthodier des Obersilurs der Insel Ösel. Ihre Paläobiologie und Paläontologie. *Palaeontographica* 76:35–94.

Hotton, N., III. 1959. The pelycosaur tympanum and early evolution of the middle ear. *Evolution* 13:99–121.

———. 1960. The chorda tympani and middle ear as guides to origin and divergence of reptiles. *Evolution* 14:194–211.

Hotton, N., III and Kitching, J. W. 1963. Speculations on Upper Beaufort deposition. *S. Afr. Jour. Sci.* 59:254–58.

Hough, J. R. 1944. The auditory region in some Miocene carnivores. *Jour. Paleont.* 18:470–79.

Howard, H. 1957. A gigantic "toothed" marine bird from the Miocene of California. *Bull. Santa Barbara Mus. Nat. Hist.* 1:1–23.

Hu, S.-Y. 1964. Carnosaurian remains from Alashan, Inner Mongolia. *Vert. Palasiatica* 8:42–55. (English summary: 56–63.)

Hubbs, C. L. 1919. The Amphibioidei, a group of fishes pròposed to include the Crossopterygii and the Dipneusti. *Science* n.s., 49:569–70.

Huene, F. 1950. Die Entstehung der Ornithischia schon früh in der Trias. *Neues Jahrb. Min. Geol. Paläont. Monatsh.* 1950(2):53–58.

———. 1952*a*. Skelett und Verwandschaft von *Simosaurus*. *Paleontographica* Abt. A 102:163–82.

———. 1952*b*. Revision der Gattung *Pleurosaurus* auf Grund neuer und alter Funde. *Palaeontographica* Abt. A 101:167–200.

———. 1959. Neues über *Tupilakosaurus*. *Neues Jahrb. Min. Geol. Paläont. Monatsh.* 1959(5):230–33.

———. 1960. Die Frage der Herkunft der Ichthyosaurier. *Neues Jahrb. Min. Geol. Paläont. Monatsh.* 1960(4):147–48.

Hürzeler, J. 1958. *Oreopithecus bambolii*, a preliminary report. *Verhandl. Naturf. Ges. Basel* 69:1–48.

———. 1959. Signification de l'Oréopithèque dans la genèse humaine. *Verhandl. Schweiz. Naturf. Ges.* 139:142–50.

Hussakof, L. 1911. The Permian fishes of North America. *Publ. Carnegie Inst. Washington* No. 146:155–75.

Inger, R. F. 1967. The development of a phylogeny of frogs. *Evolution* 21:369–84.

Jain, S. L., Robinson P. L., and Chowdhury, T. K. R. 1962. A new vertebrate fauna from the early Jurassic of the Deccan, India. *Nature* (*London*) 194(4830): 755–57.

Janensch, W. 1955. Der ornithopode *Dysalotosaurus* der Tendaguruschichten. *Palaeontographica* suppl. 7, ser. 3:105–76; cf. also suppl. 7, ser. 4:237–40.

Jarvik, E. 1950*a*. On some osteolepiform crossopterygians from the Upper Old Red Sandstone of Scotland. *Handl. K. Svenska Vetenskapsakad.* ser. 4, 2(2):1–35.

———. 1950*b*. Note on Middle Devonian crossopterygians from the eastern part of Gauss Halvö, East Greenland. *Meddel. om Grønland* 149(6):1–20.

———. 1950*c*. Middle Devonian vertebrates from Canning Land and Wegeners Halvö (East Greenland). II. Crossopterygii. *Meddel. om Grønland* 96(4):1–132.

———. 1965*a*. Die Raspelzunge der Cyclostomen und die pentadactyle

Extremität der Tetrapoden als Beweise für monophyletische Herkunft. *Zool. Anz.* 175:101–43.

———. 1965*b*. On the origin of girdles and paired fins. *Israel Jour. Zool.* 14:141–72.

———. 1965*c*. Specializations in early vertebrates. *Ann. Soc. Roy. Zool. Belgique* 94:1–95.

———. 1966. Remarks on the structure of the snout in *Megalichthys* and certain other rhipidistid crossopterygians. *Arkiv för Zool.* 19(2):41–98.

Jepsen, G. L. 1932. *Tubulodon taylori*, a Wind River Eocene tubulidentate from Wyoming. *Proc. Amer. Philos. Soc.* 71:255–74.

———. 1966. Early Eocene bat from Wyoming. *Science* 154:1333–38.

Jessen, H. 1966. Die Crossopterygier des Oberen Plattenkalkes (Devon) der Bergisch-Gladbach-Paffrather Mulde (Rheinisches Schiefergebirge) unter Berücksichtigung von amerikanischem und europäischem *Onychodus*-Material. *Arkiv för Zool.* ser. 2, 18:305–89.

Jongmans, W. J. and Gothan, W. 1937. Betrachtungen über die Ergebnisse des zweiten Kongresses für Karbonstratigraphie. *C.R. 2d Congrès pour l'Avancement des Études de Stratigraphie Carbonifère, Heerlen, 1935* 1:1–40

Kälin, J. 1955. Crocodilia. *In: Traité de Paléontologie* (ed. J. Piveteau), 5:695–784. Paris.

Karatajuté-Talimaa, V. N. 1963. [Genus *Asterolepis* from the Devonian of the Russian Platform.] *In: The Data of Geology of Lithuania. Acad. Sci. Lithuanian SSR, Vilnius*:65–224. (In Russian, English summary.)

Kermack, K. A. and Mussett, F. 1958. The jaw articulation of the Docodonta and the classification of Mesozoic mammals. *Proc. Roy. Soc. London* (*B*)148:204–15.

Kielan-Jaworowska, Z. and Kowalski, K. 1965. Polish-Mongolian palaeontological expeditions to the Gobi Desert in 1963 and 1964. *Bull. Acad. Polonaise Sci. Varsovie, Sér. Sci. Biol.* 13:175–79.

King, P. B. 1942. Permian of west Texas and southeastern New Mexico. *Bull. Amer. Assoc. Petrol. Geol.* 26:535–663.

Kjellesvig-Waering, E. N. 1961. Eurypterids of the Devonian Holland Quarry shale of Ohio. *Fieldiana: Geology* 14:79–98.

Kleinenberg, S. E. 1959. On the origin of Cetacea. *Proc. 15th Internat. Congr. Zool. London* 1958:445–47.

Konjukova, E. D. 1953. [Lower Permian terrestrial vertebrates of northern pre-Urals (Inty River basin).] *Doklady Akad. Nauk USSR* 89:723–26. (In Russian.)

Kraglievich, J. L. and Rivas, S. 1951. *Orophodon* Amegh., representate de una nueva superfamilia Orophodontoidea del suborder Xenarthra. *Comun. Inst. Invest. Cienc. Nat. Buenos Aires, Zool.* 2(2):9–28.

Krebs, B. 1963. Bau und Funktion des Tarsus eines Pseudosuchiers aus der Trias des Monte San Giorgio (Kanton Tessin, Schweiz). *Paläont. Zeitschr.* 37:88–95.

———. 1965. *Ticinosuchus ferox* nov. gen., nov. sp. Ein neuer Pseudosuchier aus der Trias des Monte San Giorgio. *Schweiz. Paläont. Abhandl.* 81:1–140.

Kretzoi, M. 1945. Bemerkungen über das Raubtiersystem. *Ann. Hist.Nat. Mus. Hung.* 38:59–83.

Kühne, W. G. 1958. Rhaetische Triconodonten aus Glamorgan, ihre Stellung zwischen den Klassen Reptilia und Mammalia und ihre Bedeutung für die Reichart'sche Theorie. *Paläont. Zeitschr.* 32:197–235.

Kuhn, O. 1958. Ein neuer Lacertilier aus dem fränkischen Lithographieschiefer. *Neues Jahrb. Min. Geol. Paleont. Monatsh.* 1958: 380–82.

———. 1960. Die Familien der fossilen Amphibien und Reptilien. *Ber. Naturforsch. Ges. Bamberg* 37:20–52.

———. 1962. Die vorzeitlichen Frösche und Salamander, ihre Gattungen und Familien. *Jahrb. Ver. Vaterl. Naturk. Württemberg* 117: 327–72.

———. 1965. *Die Amphibien.* Munich. Pp. 102.

———. 1967. Die fossile Wirbeltierklasse Pterosauria. Krailling bei München. Pp. 52.

Kuhn-Schnyder, E. 1959. Über das Gebiss von *Cyamodus.* Vierteljahrsschr. *Naturforsch. Ges. Zürich* 104: 174–88.

———. 1960. Ein neuer Pachypleurosaurier von der Stulseralp bei Bergün (Kt. Graubünden, Schweiz). *Ecl. Geol. Helv.* 52:639–58.

———. 1961. Der Schädel von *Simosaurus. Palaeont. Zeitschr.* 35:95–113.

———. 1962. Ein weiterer Schädel von *Macrocnemus bassanii* Nopcsa aus der anisischen Stufe der Trias des Monte San Giorgio (Kt. Tessin, Schweiz). *Paläont. Zeitschr. H. Schmidt-Festband*: 110–33.

———. 1963. Wege der Reptiliensystematik. *Paläont. Zeitschr.* 37:61–87.

Kulczycki, J. 1957. Upper Devonian fishes from the Holy Cross Mountains (Poland). *Acta Palaeont. Polonica* 2:285–382.

Kurtén, B. 1963. Villafranchian faunal evolution. *Comment. Biol. Soc. Sci. Fennica* 26(3):1–18.

———. 1966. Holarctic land connections in the early Tertiary. *Comment. Biol. Soc. Sci. Fennica* 29(5):1–5.

Langston, W., Jr. 1956. The Sebecosuchia: cosmopolitan crocodilians? *Amer. Jour. Sci.* 254:605–14.

———. 1960. The vertebrate fauna of the Selma formation of Alabama. Pt. VI. The dinosaurs. *Field Mus., Geol. Mem.* 3:319–61.

Langston, W., Jr. 1963. Fossil vertebrates and the late Palaeozoic Red Beds of Prince Edward Island. *Bull. Nat. Mus. Canada* No. 187 (Geol. ser. 56): 1–36.

———. 1965*a*. *Oedaleops campi* (Reptilia: Pelycosauria). A new genus and species from the Lower Permian of New Mexico, and the family Eothyrididae. *Bull. Texas Mem. Mus.* 9:1–47.

———. 1965*b*. Fossil crocodilians from Colombia and the Cenozoic history of the Crocodilia in South America. *Univ. Calif. Publ. Geol. Sci.* 52:1–157.

Lapparent, A.-F. de. 1947. Les dinosauriens du Crétacé supérieur du Midi de la France. *Mém. Soc. Géol. France* 26 (56):1–54.

———. 1955. Étude paléontologique des vertébrés du Jurassique d'El Mers (Moyen Atlas). *Notes Mém. Serv. Géol. Maroc* 124:1–36.

———. 1960. Les dinosauriens du "continental intercalaire" du Sahara Central. *Mém. Soc. Geol. France* n.s., 88*A*. Pp. 56.

Lapparent, A.-F. de and Lavocat, R. 1955. Dinosauriens. *In*: *Traité de Paléontologie* (ed. J. Piveteau), 5:785–962. Paris.

Lapparent, A.-F. de and Zbyszewski, G. 1957. Les dinosauriens de Portugal. *Mém. Serv. Géol. Portugal* n.s., 2:1–63.

Leakey, L. S. B. 1962. A new Lower Pliocene fossil primate from Kenya. *Ann. Mag. Nat. Hist.* ser. 13, 4:689–96.

———. 1966. Africa and Pleistocene overkill? *Nature* (*London*) 212:1615–16.

Leakey, L. S. B., Tobias, P. V., and Napier, J. R. 1964. A new species of the genus *Homo* from Olduvai Gorge. *Nature* (*London*) 202:7–9.

LeGros Clark, W. E. 1934. *Early Forerunners of Man*. London and Baltimore. Pp. 296.

LeGros Clark, W. E. and Thomas, D. P. 1951. Associated jaws and limb bones of *Limnopithecus macinnesi*. *Fossil Mammals of Africa No. 4, Brit. Mus.* (*Nat. Hist.*). London. Pp. 27.

Lehman, J.-P. 1955. Phyllospondyli. *In*: *Traité de Paléontologie* (ed. J. Piveteau), 5:227–49. Paris.

———. 1956. Compléments à l'étude des genres *Ecrinesomus* et *Bobasatrania* de l'Eotrias de Madagascar. *Ann. Paléont.* 42:65–94.

———. 1962. A propos de la double articulation de la cuirasse des Arthrodires. *In*: Problèmes Actuels de Paléontologie (Évolution des Vertébrés). *Coll. Internat. Centre Nat. Recher. Sci.* No. 104:63–68.

Leonardi, A. 1959. L'ittiofauna del "tripoli" del Miocene superiore di Bessima (Enna). *Palaeontogr. Ital.* 54:115–73.

Lewis, G. E. 1934. Preliminary notice of new man-like apes from India. *Amer. Jour. Sci.* ser. 5, 27: 161–79.

Lewis, G. E. and Vaughn, P. P. 1965. Early Permian vertebrates from the Cutler Formation of the Placerville area, Colorado. *U.S. Geol. Surv. Prof. Paper* 503-C:1–50.

Lindström, M. 1964. Conodonts. Amsterdam. Pp. 196.

Liu, H. T. and Chang, M. M. 1963. The first discovery of a helicoprionid from China. *Vert. Palasiatica* 7: 123–31.

Liu, H. T. and others. 1963. Lycopterid fishes from North China. *Mem. Inst. Vert. Paleont. Peking* 6:1–53.

Liu, T. S. and P'an, K. 1958. Devonian fishes from Wutung series near Nanking, China. *Palaeont. Sinica* ser. C, 15:41 pp.

Liu, Y.-H. 1963. On the Antiarchi from Chutsing, Yunnan. *Vert. Palasiatica* 7:39–47.

———. 1965. New Devonian agnathans of Yunnan. *Vert. Palasiatica* 9:125–34.

Lund, R. 1966. Intermuscular bones in *Pholidophorus bechei* from the Lower Lias of England. *Science* 152:348–49.

McDowell, S. B., Jr. 1948. The bony palate of birds. Part I. The Palaeognathae. *The Auk* 65:520–49.

———. 1958. The Greater Antillean insectivores. *Bull. Amer. Mus. Nat. Hist.* 115:113–214.

McDowell, S. B., Jr. and Bogert, C. M. 1954. The systematic position of *Lanthanotus* and the affinities of the anguinomorphan lizards. *Bull. Amer. Mus. Nat. Hist.* 105:1–142.

McGrew, P. O. and Patterson, B. 1962. A picrodontid insectivore (?) from the Paleocene of Wyoming. *Breviora, Mus. Comp. Zool.* No. 175:1–9.

MacIntyre, G. T. 1966. The Miacidae (Mammalia, Carnivora). Part I. The systematics of *Ictidopappus* and *Proictis*. *Bull. Amer. Mus. Nat. Hist.* 131:115–210.

McKenna, M. C. 1956. Survival of primitive notoungulates and condylarths into the Miocene of Colombia. *Amer. Jour. Sci.* 254:736–43.

———. 1960. Insectivora fossils. *McGraw-Hill Encyclopedia of Science and Technology* 8:143–44.

———. 1962. Collecting small fossils by washing and screening. *Curator* 3:221–35.

———. 1963. New evidence against tupaioid affinities of the mammalian family Anagalidae. *Amer. Mus. Novit.* No. 2158:1–16.

———. 1965*a*. Collecting microvertebrate fossils by washing and screening. *In*: *Handbook of Paleontological Techniques* (eds. B. Kummel and D. Raup). San Francisco. Pp. 193–203.

———. 1965*b*. Paleontology and the origin of the primates. *Folia Primat.* 4:1–25.

McLaren, I. A. 1960. Are the Pinnipedia biphyletic? *Syst. Zool.* 9:18–28.

Maleev, E. A. 1952. [A new ankylosaur from the Upper Cretaceous of Mongolia.] *Doklady Akad. Nauk URSS* n.s. 87:273–76. (In Russian.)

Maleev, E. A. 1954. [Armored dinosaurs of the Upper Cretaceous of Mongolia.] *Trav. Inst. Paléozool. Akad. Nauk SSSR* 48: 142–70. (In Russian.)

———. 1955*a*. [New carnivorous dinosaurs from the Upper Cretaceous of Mongolia.] *Doklady Akad. Nauk URSS* 104: 779–82. (In Russian.)

———. 1955*b*. [A gigantic dinosaur from Mongolia.] *Doklady Akad. Nauk URSS* 104:634–37. (In Russian.)

Malzahn, E. 1963. *Lepidotus elvensis* Blainville aus dem Posidonienschiefer der Dobbertiner Liasscholle. *Geol. Jahrb.* 80:539–60.

Marples, B. J. 1952. Early Tertiary penguins of New Zealand. *New Zealand Geol. Surv., Paleont. Bull.* 20:1–66.

Martin, P. S. 1966. Africa and Pleistocene overkill. *Nature* (*London*) 212:339–42.

Matthew, W. D. and Granger, W. 1925. Fauna and correlation of the Gashato Formation of Mongolia. *Amer. Mus. Novit.* No. 189:1–12.

Mayr, E. 1963. The taxonomic evaluation of fossil hominids. *In*: *Classification and Human Evolution* (ed. S. L. Washburn). Chicago. Pp. 332–46.

Meise, W. 1963. Verhalten der Straussartigen Vögel und Monophylie der Ratitae. *Proc. XIII Internat. Ornith. Congr.* 1:115–25.

Meszoely, C. 1966. North American fossil cryptobranchid salamanders. *Amer. Midl. Nat.* 75:495–515.

Miles, R. S. and Westoll, T. S. 1963. Two new genera of coccosteid Arthrodira from the middle Old Red Sandstone of Scotland, and their stratigraphical distribution. *Trans. Roy. Soc. Edinburgh* 65:179–210.

Miller, A. K., Cullison, J. S., and Youngquist, W. 1947. Lower Ordovician fish remains from Missouri. *Amer. Jour. Sci.* 245:31–34.

Moodie, R. L. 1916. The Coal Measures Amphibia of North America. *Publ. Carnegie Inst. Washington* No. 238. Pp. 222.

Moss, M. L. 1964. The phylogeny of mineralized tissues. *In*: *International Review of General and Experimental Zoology* 1:298–332.

Napier, J. R. and Davis, P. R. 1959. The forelimb skeleton and associated remains of *Proconsul africanus*. *Fossil Mammals of Africa No. 16. Brit. Mus.* (*Nat. Hist.*). London. Pp. 69.

Nevo, E. 1968. Pipid frogs from the early Cretaceous of Israel and pipid evolution.Bull. Mus. Comp. Zool. 136:255–318.

Nielsen, E. 1949. Studies on Triassic fishes from East Greenland. II. *Australosomus* and *Birgeria*. *Palaeozool. Groenland.* Pp. 309.

———. 1952. On new or little known Edestidae from the Permian and Triassic of East Greenland. *Meddel. om Grønland.* 144(5):1–55.

———. 1954. *Tupilakosaurus heilmani* n.g. et n. sp. An interesting batrachomorph from the Triassic of East Greenland. *Meddel. om Grønland* 72(8):1–33.

Noble, G. K. 1931. *The Biology of the Amphibia*. New York. Pp. 577.

Nursall, J. R. 1962. Swimming and the origin of paired appendages. *Amer. Zool.* 2:127–41.

Nybelin, O. 1961*a*. Über die Frage der Abstammung der rezenten primitiven Teleostier. *Paläont. Zeitschr.* 35:114–17.

———. 1961*b*. *Leptolepis dubia* aus den Torleiten-Schichten des Oberen Jura von Eichstätt. *Paläont. Zeitschr.* 35:118–22.

———. 1963. Zur Morphologie und Terminologie des Schwanzskelettes der Actinopterygier. *Arkiv för Zool.* ser. 2, 15:485–516.

———. 1964. Versuch einer taxonomischen Revision der jurassischen Fischgattung *Thrissops* Agassiz. *Göteborgs Kungl. Vetenskaps-Och Vitterhets-Samhälles, Handl. 6 Följ.* ser. B, 9(4):1–44.

———. 1966. On certain Triassic and Liassic representatives of the family Pholidophoridae S. Str. *Bull. Brit. Mus. (Nat. Hist.), Geol.* 11(8):353–432.

Oakley, K. P. (ed.). 1950. The Pliocene-Pleistocene boundary. *Rept. 18th Internat. Geol. Congr., London, 1948*. Part 9. London. Pp. 130.

Obruchev, D. V. 1945. The evolution of Agnatha. *Zool. Zhurn. Moscow* 24:257–72. (In Russian, English summary.)

———. 1953. [The study of the edestids and the work of A. P. Karpinsky.] *Trudy Paleont. Inst. Akad. Nauk SSSR* 45:1–86 (In Russian.)

———. 1964. Agnatha, Pisces. *In*: *Osnovy Paleontologii* (ed. J. A. Orlov), 11:34–173. Moscow. (In Russian.)

Obrucheva, O. P. 1962. [Placodermi from the Devonian of USSR. Coccosteidae and Dinichthyidae.] Moscow Univ. Pp. 189. (In Russian.)

Oliver, W. R. B. 1949. The moas of New Zealand and Australia. *Bull. Dominion Mus.* 15:1–206.

Olson, E. C. 1947. The family Diadectidae and its bearing on the classification of reptiles. *Fieldiana: Geology* 11:1–53.

———. 1951. Fauna of Upper Vale and Choza: 2. A new captorhinomorph reptile. *Fieldiana: Geology* 10:97–104.

———. 1965. New Permian vertebrates from the Chickasha Formation in Oklahoma. *Okla. Geol. Surv. Circular* No. 70:1–70.

———. 1966. Relationships of *Diadectes*. *Fieldiana: Geology* 14:199–227.

Orlov, J. A. 1933. *Semantor macrurus* (ordo Pinnipedia, fam. Semantoridae fam. nov.) aus den Neogen-Ablagerungen Westsibiriens. *Trav. Inst. Paléozool. Acad. Sci. URSS* 2:165–268.

Ørvig, T. 1960. New finds of acanthodians, arthrodires, crossopterygians, ganoids and dipnoans in the upper Middle Devonian calcareous flags (Oberer Plattenkalk) of the Bergisch Gladbach-Paffrath trough (Part 1). *Paläont. Zeitschr.* 34:295–335.

———. 1965*a*. Palaeohistological notes. 2. Certain comments on the phyletic significance of acellular bone tissue in early lower vertebrates. *Arkiv Zool.* 16:551–56.

———. 1965*b*. Phylogeny of tooth tissues. *In*: *Structural and Chemical Organization of Teeth* (eds. A. E. Miles and R. C. Grenlach). New York. Pp. 45–105.

———. 1966. Histologic studies of ostracoderms, placoderms, and fossil elasmobranchs. 2. On the dermal skeleton of two late Paleozoic elasmobranchs. *Arkiv Zool.* 19: 1–39.

Osborn, H. F. 1907. *Evolution of Mammalian Molar Teeth.* New York. Pp. 250.

Ostrom, J. H. 1962. The cranial crests of hadrosaurian dinosaurs. *Postilla, Peabody Mus. Yale Univ.* 62:1–29.

Panchen, A. L. 1963. The homologies of the labyrinthodont centrum. *Proc. 16th Internat. Congr. Zool. Washington* 1:161.

———. 1964. The cranial anatomy of two Coal Measure anthracosaurs. *Philos. Trans. Roy. Soc. London* (*B*) 247:593–637.

———. 1966. The axial skeleton of the labyrinthodont *Eogyrinus attheyi*. *Jour. Zool. London* 150:199–222.

Panchen, A. L. and Walker, A. D. 1960. British Coal Measure labyrinthodont localities. *Ann. Mag. Nat. Hist.* ser. 13, 3: 321–32.

Parrington, F. R. 1950. The skull of *Dipterus*. *Ann. Mag. Nat. Hist.* ser. 12, 3:534–47.

———. 1956. The patterns of dermal bones in primitive vertebrates. *Proc. Zool. Soc. London* 127:389–411.

———. 1958. The problem of the classification of reptiles. *Jour. Linn. Soc. London* (*Zool.*) 44:99–115.

Parsons, T. S. and Williams, E. E. 1961. Two Jurassic turtle skulls; a morphological study. *Bull. Mus. Comp. Zool.* 125:43–107.

Patterson, B. 1952. Un nuevo y extraordinario marsupial deseadiano. *Rev. Mus. Munic. Cienc. Nat. Mar del Plata* 1:39–44.

———. 1954. The geologic history of non-hominid primates in the Old World. *Human Biology* 26:191–209.

———. 1957. Mammalian phylogeny. *In*: *Premier symposium sur la spécificité parasitaire des parasites des vertébrés.* Neuchatel. Pp. 15–49.

———. 1962. An extinct solenodontid insectivore from Hispaniola. *Breviora, Mus. Comp. Zool.* 165:1–11.

Patterson, B. and McGrew, P. O. 1962. A new arctocyonid from the Paleocene of Wyoming. *Breviora, Mus. Comp. Zool.* 174:1–10.

Patterson, B. and Olson, E. C. 1962. A triconodontid mammal from the Triassic of Yunnan. Internat. Coll. on the Evolution of Mammals. *Kon. Vlaamse Acad. Wetensch. Lett. Sch. Kunsten België, Brussels, 1961* 1:129–91.

Paula Couto, C. de. 1952*a*. Fossil mammals from the beginning of the Cenozoic in Brazil. Marsupialia: Didelphidae. *Amer. Mus. Novit.* No. 1567:1–26.

———. 1952*b*. Fossil mammals from the beginning of the Cenozoic in Brazil. Marsupialia: Polydolopidae and Borhyaenidae. *Amer. Mus. Novit.* No. 1559: 1–27.

Peabody, F. E. 1952. *Petrolacosaurus kansensis* Lane, a Pennsylvanian reptile from Kansas. *Vertebrata, Art. I, Paleont. Contrib. Univ. Kansas* No. 10: 1–41.

———. 1958. An embolomerous amphibian in the Garnett fauna (Pennsylvanian) of Kansas. *Jour. Paleont.* 32:571–73.

Peach, B. N. and Horne, J. 1899. The Silurian rocks of Britain. Vol. 1, Scotland. *Gen. Mem. Geol. Surv. U.K.* Pp. 749.

Petronievics, B. 1950. Les deux oiseaux fossiles les plus anciens (*Archaeopteryx* et *Archaeornis*). *Ann. Géol. Pén. Balkan* 18:89–127.

Peyer, B. 1956. Über Zähne von Haramiyiden, von Triconodonten und von wahrscheinlich synapsiden Reptilien aus dem Rhät von Hallau, Kt. Schaffhausen, Schweiz. *Schweiz. Paläont. Abhandl.* 72:1–72.

Peyer, B. and Kuhn-Schnyder, E. 1955. Squamates du Trias. *In*: *Traité de Paléontologie* (ed. J. Piveteau), 5:578–605. Paris.

Price, L. I. 1937. Two new cotylosaurs from the Permian of Texas. *Proc. New England Zool. Club* 16:97–102.

———. 1950. Os crocodilideos da fauna da Formação Baurú, do Cretáceo terrestre do Brasil Meridional. *An. Acad. Brasil, Cien.* 22:473–90.

———. 1955. Novos crocodilideos dos arenitos da Série Baurú, Cretáceo do Estado de Minas Gerais. *An. Acad. Brasil, Cien.* 27:487–98.

Quinn, J. H. 1955. Miocene Equidae of the Texas Gulf Coastal Plain. *Univ. Texas Publ.* No. 5516:1–102.

Radinsky, L. 1961. Tooth histology as a taxonomic criterion for cartilaginous fishes. *Jour. Morph.* 109:73–92.

———. 1966*a*. The families of the Rhinocerotoidea (Mammalia, Perissodactyla). *Jour. Mammal.* 47:631–39.

———. 1966*b*. The adaptive radiation of the phenacodontid condylarths and the origin of the Perissodactyla. *Evolution* 20:408–17.

Reed, C. A. 1960. Polyphyletic or monophyletic ancestry of mammals, or: what is a class? *Evolution* 14:314–22.

Reig, O. 1959. Primeros datos descriptivos sobre nuevos reptiles arcosaurios del Triasico de Ischigualasto (San Juan, Argentina). *Rev. Assoc. Geol. Arg.* 13:257–70.

Reig, O. 1961. Acerca da la posición sistemática de la familia Rauisuchidae y del género *Saurosuchus* (Reptilia, Thecodontia). *Publ. Mus. Munic. Cienc. Nat. Mar del Plata* 1 (3): 73–114.

———. 1963. La presencia de dinosaurios saurisquios en los "estratos de Ischigualasto" (Mesotriásico Superior) de las provincias de San Juan y La Rioja (República Argentina). *Ameghiniana* 3:3–20.

———. 1967. Archosaurian reptiles: A new hypothesis on their origins. *Science 157:565–68.*

Rhodes, F. H. T. 1954. The zoological affinities of the conodonts. *Biol. Rev.* 29:419–52.

Riabinin, A. N. 1948. [A note about a flying reptile from the Kara-tan Jurassic.] *Trav. Inst. Paleont. Akad. Nauk URSS* 15:86–93. (In Russian.)

Ride, W. D. L. 1964. A review of Australian fossil marsupials. *Jour. Roy. Soc. W. Australia* 47:97–131.

Robertson, J. D. 1957. The habitat of the early vertebrates. *Biol. Rev.* 32:156–87.

Robinson, J. T. 1963. Adaptive radiation in the australopithecines and the origin of man. *In*: *African Ecology and Human Evolution* (eds. F. C. Howell and F. Bourlière). Chicago. Pp. 385–416.

———. 1966. The distinctiveness of *Homo habilis*. *Nature* (*London*) 209: 957–60.

———. 1967. Variation and the taxonomy of the early hominids. *In*: *Evolutionary Biology* (eds. T. Dobzhansky, M. K. Hecht, W. C. Steere), vol. 1: 69–100. New York.

Rolfe, W. D. I. 1961. The geology of the Hagshaw Hills Silurian inlier, Lanarkshire. *Proc. Geol. Soc. London* No. 1585:48–52.

Romer, A. S. 1927. The pelvic musculature of ornithischian dinosaurs. *Acta Zool.* 8:225–75.

———. 1930. The Pennsylvanian tetrapods of Linton, Ohio. *Bull. Amer. Mus. Nat. Hist.* 59:77–147.

———. 1933. Eurypterid influence on vertebrate history. *Science* n.s., 78:114–17.

———. 1936. The dipnoan cranial roof. *Amer. Jour. Sci.* ser. 5, 32:241–56.

———. 1941. Notes on the crossopterygian hyomandibular and braincase. *Jour. Morphol.* 69(1):141–60.

———. 1942. Cartilage an embryonic adaptation. *Amer. Nat.* 76:394–404.

———. 1945. The late Carboniferous vertebrate fauna of Kounova (Bohemia) compared with that of the Texas redbeds. *Amer. Jour. Sci.* 243:417–42.

———. 1947. The relationships of the Permian reptile *Protorosaurus*. *Amer. Jour. Sci.* 245:19–30.

Romer, A. S. 1948. Ichthyosaur ancestors. *Amer. Jour. Sci.* 246:109–21.
———. 1952. Late Pennsylvanian and early Permian vertebrates of the Pittsburgh–West Virginia region. *Ann. Carnegie Mus.* 33:47–112.
———. 1955*a*. Herpetichthyes, Amphibioidei, Choanichthyes or Sarcopterygii? *Nature* (*London*) 176:126.
———. 1955*b*. The primitive vertebrate as a dual animal—somatic and visceral. *Proc. Zool. Soc. London* 125:811.
———. 1957. Origin of the amniote egg. *Sci. Monthly* 85 (2):57–63.
———. 1958. Tetrapod limbs and early tetrapod life. *Evolution* 12:365–69.
———. 1959. A mounted skeleton of the giant plesiosaur *Kronosaurus*. *Breviora, Mus. Comp. Zool.* No. 112:1–15.
———. 1960. Vertebrate-bearing continental Triassic strata in Mendoza region, Argentina. *Bull. Geol. Soc. Amer.* 71:1270–94.
———. 1962*a*. Vertebrate evolution. Reviews of: J.-P. Lehman, *L'Évolution des Vertébrés Inferieurs*, and E. Jarvik, *Théories de l'Évolution des Vertébrés*. *Copeia* 1962:223–27.
———. 1962*b*. The fossiliferous Triassic deposits of Ischigualasto, Argentina. *Breviora, Mus. Comp. Zool.* No. 156:1–7.
———. 1962*c*. La evolución explosiva de los rhynchosaurios del Triasico. *Rev. Mus. Arg. Sci. Nat. "Bernardino Rivadavia"* 8:1–14.
———. 1963. The "ancient history" of bone. *Ann. N.Y. Acad. Sci.* 109:168–76.
———. 1964. Bone in early vertebrates. *In*: *Bone Biodynamics* (ed. H. M. Frost). Boston. Pp. 13–40.
———. 1966. The Chañares (Argentina) Triassic reptile fauna. I. Introduction. *Breviora, Mus. Comp. Zool.* No. 247:1–14.
———. 1967*a*. The Chañares (Argentina) Triassic reptile fauna. III. Two new gomphodonts, *Massetognathus pascuali* and *M. teruggii*. *Breviora, Mus. Comp. Zool.* No. 264:1–25.
———. 1967*b*. Early reptilian evolution re-viewed. *Evolution.* 21:821–33.
Romer, A. S. and Grove, B. H. 1935. Environment of the early vertebrates. *Amer. Midl. Nat.* 16:805–56.
Romer, A. S. and Jensen, J. A. 1966. The Chañares (Argentina) Triassic reptile fauna. II. Sketch of the geology of the Rio Chañares—Rio Gualo region. *Breviora, Mus. Comp. Zool.* No. 252:1–20.
Romer, A. S. and Olson, E. C. 1954. Aestivation in a Permian lungfish. *Breviora, Mus. Comp. Zool.* No. 30:1–8.
Rosen, D. E. 1964. The relationships and taxonomic position of the halfbeaks, killifishes, silversides, and their relatives. *Bull. Amer. Mus. Nat. Hist.* 127:217–68.
Rozhdestvenskii, A. K. 1952*a*. [A new dinosaur from the Upper

Cretaceous deposits of Mongolia.] *Doklady Akad. Nauk URSS* 86:405–8. (In Russian.)

———. 1952*b*. [Discovery of *Iguanodon* in Mongolia.] *Doklady Akad. Nauk URSS* 84:1243–46. (In Russian.)

Rozhdestvenskii, A. K. and Tatarinov, L. P. 1964. *In*: *Osnovy Paleontologii*, vol. 5 (ed. J. A. Orlov). [Amphibians, reptiles, and birds.] Moscow. Pp. 722. (In Russian.)

Rusconi, C. 1951. Laberintodontes Triasicos y Pérmicos de Mendoza. *Rev. Mus. Hist. Nat. Mendoza* 5:33–158.

Russell, D. E. and McKenna, M. C. 1961. Étude de *Paroxyclaenus*, mammifère des phosphorites du Quercy. *Bull. Soc. Géol. France* ser. 7, 3:274–82.

Russell, L. S. 1948. The dentary of *Troödon*, a genus of theropod dinosaurs. *Jour. Paleont.* 22:625–29.

Säve-Söderbergh, G. 1932. Preliminary note on Devonian stegocephalians from East Greenland. *Meddel. om Grønland* 94(7):1–107.

———. 1934. Some points of view concerning the evolution of the vertebrates and the classification of this group. *Arkiv. Zool.* 26*A*(17):1–20.

———. 1937. On *Rhynchodipterus elginensis* n.g., n.sp., representing a new group of dipnoan-like Choanata from the Upper Devonian of East Greenland and Scotland. *Arch. Zool.* 29*B*:1–8.

Schaeffer, B. 1952*a*. Rates of evolution in the coelacanth and dipnoan fishes. *Evolution* 6:101–11.

———. 1952*b*. The Triassic coelacanth fish *Diplurus* with observations on the evolution of the Coelacanthini. *Bull. Amer. Mus. Nat. Hist.* 99:25–78.

———. 1963. Cretaceous fishes from Bolivia, with comments on pristid evolution. *Amer. Mus. Novit.* No. 2159:1–20.

Schaub, S. 1958. Simplicidentata. *In*: *Traité de Paléontologie* (ed. J. Piveteau), 6(2):659–818. Paris.

Schmalhausen, I. I. 1959. The origin of the Amphibia. *Proc. 15th Internat. Congr. Zool., London, 1958* Sect. 5 (5):455–58.

Schmidt, M. 1928. *Die Lebewelt unser Trias*. Öhringen. Pp. 461; supplement, 1938. Pp. 143.

Scourfield, D. J. 1937. An anomalous fossil organism, possibly a new type of chordate, from the Upper Silurian of Lesmahagow, Lanarkshire—*Ainiktozoon loganense*, gen. et sp. nov. *Proc. Roy. Soc. London* (B)121:533–47.

Sera, G. L. 1954. Posizione zoologica dei Pyrotherii. *Monit. Zool. Ital.* 62:42–44.

Shikama, T. 1947. *Teilhardosaurus* and *Endotherium*, new Jurassic Reptilia and Mammalia from the Husin coalfield, South Manchuria. *Proc. Japan Acad.* 23:76–84.

Shikama, T. 1966. Postcranial skeletons of Japanese *Desmostylia. Palaeont. Soc. Japan* Spec. Pap. No. 12:1–202.
Shishkin, M. A. 1961. [New data on *Tupilakosaurus.*] *Doklady Akad. Nauk USSR* 136:938–41. (In Russian.)
Sigogneau, D. 1963. Remarks on Gorgonopsia. *S. Afr. Jour. Sci.* 59:207–9.
Sill, W. D. 1967. *Proterochampsa barrionuevoi* and the early evolution of the Crocodilia. *Bull. Mus. Comp. Zool.* 135:415–46.
Simmons, D. J. 1965. The non-therapsid reptiles of the Lufeng Basin, Yunnan, China. *Fieldiana: Geology* 15:1–93.
Simonetta, A. M. 1961. On the mechanical implications of the avian skull and their bearing on the evolution and classification of birds. *Quart. Rev. Biol.* 35:206–20.
Simons, E. L. 1959. An anthropoid frontal bone from the Fayûm Oligocene of Egypt: the oldest skull fragment of a higher primate. *Amer. Mus. Novit.* No. 1976:1–16.
———. 1960. New fossil primates: a review of the past decade. *Amer. Scient.* 48:179–92.
———. 1962. Two new primate species from the African Oligocene. *Postilla, Peabody Mus. Yale Univ.* No. 64:1–12.
———. 1964. On the mandible of *Ramapithecus. Proc. Nat. Acad. Sci.* 51:528–35.
Simpson, G. G. 1928. Further notes on Mongolian Cretaceous mammals. *Amer. Mus. Novit.* No. 329:1–14.
———. 1947. *Haramiya,* new name, replacing *Microcleptes* Simpson, 1928. *Jour. Paleont.* 21:497.
———. 1948. The Eocene of the San Juan Basin, New Mexico. *Amer. Jour. Sci.* 246:257–82; 363–85.
———. 1953. *The Major Features of Evolution.* New York. Pp. 434.
———. 1959. Mesozoic mammals and the polyphyletic evolution of mammals. *Evolution* 13:405–14.
———. 1960*a*. The nature and origin of supraspecific taxa. *Cold Spring Harbor Symposia on Quantitative Biology* 24:255–71.
———. 1960*b*. Diagnosis of the classes Reptilia and Mammalia. *Evolution* 14:388–92.
———. 1961. Historical zoogeography of Australian mammals. *Evolution* 15:431–46.
———. 1967. The Tertiary lorisiform primates of Africa. *Bull. Mus. Comp. Zool.* 136(3):39–62.
Slaughter, B. H. 1965. A therian from the Lower Cretaceous (Albian) of Texas. *Postilla, Peabody Mus. Yale Univ.* No. 93:1–18.
Smith, I. C. 1956. A note on the axial skeleton of the anaspid *Pharyngolepis* sp. *Arkiv. Zool.* ser. 2, 9:573–78.
———. 1957. New restorations of the heads of *Pharyngolepis oblongus*

Kiaer and *Pharyngolepis kiaeri* sp. nov., with a note on their lateral-line systems. *Norsk. Geol. Tidsskr.* 37:373–402.

Squirrel, H. C. 1958. New occurrences of fish remains in the Silurian of the Welsh Borderland. *Geol. Mag.* 95:328–32.

Steen, M. 1931. The British Museum collection of Amphibia from the Middle Coal Measures of Linton, Ohio. *Proc. Zool. Soc. London* 1930:849–91.

———. 1934. The amphibian fauna from the South Joggins, Nova Scotia. *Proc. Zool. Soc. London.* Pp. 465–504.

Stensiö, E. 1932. Triassic fishes from East Greenland. *Meddel. om Grønland* 83(3):1–305.

———. 1937. Notes on the endocranium of a Devonian *Cladodus. Bull. Geol. Inst. Upsala* 27:128–44.

———. 1947. The sensory lines and dermal bones of the cheek in fishes and amphibians. *K. Svenska Vetenskapsakad. Handl.* ser. 3, 24 (3):1–195.

———. 1958. Les cyclostomes fossiles ou ostracodermes. *In*: *Traité de Zoologie* (ed. P. P. Grassé), 13 (1):173–425. Paris.

———. 1962. Origine et nature des écailles placoïdes et des dents. *In*: Problemes Actuels de Paléontologie (Evolution des Vertébrés). *Coll. Internat. Cent. Nat. Recher. Sci.* No. 104:75–85.

———. 1963. The brain and the cranial nerves in fossil, lower craniate vertebrates. *Skr. Norske Videnskaps. Akad. Oslo, Mat.-Naturv. Kl.* n.s., 13:1–120.

Sternberg, C. M. 1945. Pachycephalosauridae proposed for dome-headed dinosaurs, *Stegoceras lambei*, n. sp., described. *Jour. Paleont.* 19:534–38.

———. 1949. The Edmonton fauna and description of a new *Triceratops* from the Upper Edmonton member; phylogeny of the Ceratopsidae. *Bull. Nat. Mus. Canada* 113:33–46.

———. 1953. A new hadrosaur from the Oldman Formation of Alberta: discussion of nomenclature. *Bull. Nat. Mus. Canada* 128:275–86.

Stirton, R. A. 1951. Ceboid monkeys from the Miocene of Colombia. *Univ. Calif. Publ. Bull. Dept. Geol. Sci.* 28:315–56.

———. 1953. Vertebrate paleontology and continental stratigraphy in Colombia. *Bull. Geol. Soc. Amer.* 64:603–22.

Stirton, R. A. and Savage, D. E. 1950. A new monkey from the La Venta Miocene of Colombia. *Compil. Estud. Geol. Ofic. Colombia* 8:345–56.

Stovall, J. W. and Langston, W., Jr. 1950. *Acrocanthosaurus atokensis*, a new genus and species of Lower Cretaceous Theropoda from Oklahoma. *Amer. Midl. Nat.* 43:696–728.

Stovall, J. W., Price, L. I., and Romer, A. S. 1966. The postcranial skeleton of the giant Permian pelycosaur *Cotylorhynchus romeri*. *Bull. Mus. Comp. Zool.* 135:1–30.

Tabaste, N. 1964. Étude de restes de poissons du Crétacé Saharien. *Mém. Inst. Franç. Afr. Noire* No. 86:437–85.

Tarlo, L. B. H. 1960. A review of Upper Jurassic pliosaurs. *Bull. Brit. Mus. (Nat. Hist.) Geol.* 4:145–89.

———. 1961. *Rhinopteraspis cornubica* (McCoy) with notes on the classification and evolution of the pteraspids. *Acta Palaeont. Polonica* 6:367–402.

———. 1962. Lignées évolutives chez les Ostracodermes hétérostracés. *In*: Problèmes Actuels de Paléontologie (Evolution des Vertébrés). *Colloq. Internat. Centre Nat. Recher. Sci.* No. 104:31–37.

———. 1963. A primitive whale from the London Clay of the Isle of Sheppy. *Proc. Geol. Assoc.* 74:319–23.

Tatarinov, L. P. 1959. [The origin of reptiles and some principles of their classification.] *Paleont. Zhurn.* No. 4:65–84. (In Russian.)

———. 1964. Lepidosauria. *In*: *Osnovy Paleontologii* (ed. J. A. Orlov), 5 [Amphibia, Reptiles, Birds]: 439–92. Moscow. (In Russian.)

Tate, G. H. H. 1948. Results of the Archbold Expeditions. No. 60. Studies in the Peramelidae (Marsupialia). *Bull. Amer. Mus. Nat. Hist.* 92:313–46.

Thaler, L. 1965. Une échelle de zones biochronologiques pour les mammifères du Tertiaire d'Europe. *C. R. Sommaire Séances Soc. Géol. France* 1965 (4):118.

———. 1966. Les rongeurs fossiles du Bas-Languedoc dans leurs rapports avec l'histoire des faunes et la stratigraphie du Tertiaire d'Europe. *Mém. Mus. Natl. Hist. Nat.*, n. ser. C, 17:1–295.

Thenius, E. 1949. Die Carnivoren von Göriach (Steiermark). Beiträge zur Kenntnis der Säugetierreste des steirischen Tertiärs IV. *Sitzungsber. Österr. Akad. Wiss. Math.-Nat. Kl.* Abt. 1, 158:695–762.

———. 1959. Ursidenphylogenese und Biostratigraphie. *Zeitschr. Säugetierk.* 24:78–84.

Thorsteinsson, R. 1958. Cornwallis and Little Cornwallis islands, District of Franklin, Northwest Territories. *Mem. Geol. Surv. Canada* 294:1–134.

Tobien, H. 1963. Zur Gebisz-Entwicklung tertiärer Lagomorphen (Mamm.). *Europas. Notizbl. Hess. Landesamt. Bodenforsch.* 91:16–35.

Trewavas, E., White, E. I., Marshall, N. B., and Tucker, D. W. 1955. [Discussion of "Herpetichthyes, Amphibioidei, Choanichthyes or Sarcopterygii?" by A. S. Romer.] *Nature (London)* 176:126–27.

Turnbull, W. D. and Turnbull, P. F. 1955. A recently discovered *Phlegethontia* from Illinois. *Fieldiana: Zoology* 37:523–35.

Underwood, G. 1954. On the classification and evolution of geckos. *Proc. Zool. Soc. London* 124:469–92.

———. 1967. A contribution to the classification of snakes. *Brit. Mus. (Nat. Hist.) Publ.* 653:1–179.

Van Valen, L. 1960. Therapsids as mammals. *Evolution* 14:304–13.

———. 1964. A possible origin for rabbits. *Evolution* 18:484–91.

———. 1965*a*. A middle Paleocene primate. *Nature (London)* 207:435–36.

———. 1965*b*. Some European Proviverrini (Mammalia, Deltatheridia). *Paleontology* 8:638–65.

———. 1965*c*. Treeshrews, primates, and fossils. *Evolution* 19:137–51.

———. 1965*d*. Paroxyclaenidae, an extinct family of Eurasian mammals. *Jour. Mammal.* 46:388–97.

Van Valen, L. and Sloan, R. E. 1965. The earliest primates. *Science* 150:743–45.

Vaughn, P. P. 1958. On a new pelycosaur from the Lower Permian of Oklahoma, and on the origin of the family Caseidae. *Jour. Paleont.* 32:981–91.

———. 1960. On the possibly polyphyletic origin of reptiles. *Evolution* 14:274–76.

———. 1962. The Paleozoic microsaurs as close relatives of reptiles, again. *Amer. Midl. Nat.* 67:79–84.

———. 1964. Vertebrates from the Organ Rock shale of the Cutler Group, Permian of Monument Valley and vicinity, Utah and Arizona. *Jour. Paleont.* 38:567–83.

Viret, J. 1949. Sur le *Pliohyrax rossignoli* du Pontien de Soblay (Aïn). *C.R. Acad. Sci. Paris* 228: 1742–44.

———. 1951. Catalogue critique de la faune des mammifères Miocènes de la Grive Saint-Alban (Isère). Part I. *Nouv. Arch. Mus. Hist. Nat. Lyon.* 3:1–104.

———. 1961. Artiodactyla. *In*: *Traité de Paléontologie* (ed. J. Piveteau), 6(1):886–1084. Paris.

Vorobjeva, E. I. 1959. [A new genus of crossopterygian fish *Platycephalichthys* from the Upper Devonian of the River Lovat.] *Palaeont. Zhurn. Akad. Nauk USSR.* 3:95–106.

———. 1960. [New facts about the genus *Panderichthys* from the Devonian of the U.S.S.R.] *Paläont. Zhurn. Akad. Nauk USSR* 1960:87–96. (In Russian.)

———. 1962. [Rhizodont crossopterygian fishes of the principal Devonian regions of the SSSR.] *Trudy Paleont. Inst. Akad. Nauk URSS* 94:1–139. (In Russian.)

Vyushkov, B. P. 1957. [New kotlassiomorphs from the Tatar deposits of European SSSR.] *Trudy Paleont. Inst. Akad. Nauk SSSR* 68:89–107. (In Russian.)

Vyuskhov, B. P. and Chudinov, P. K. 1967. [New captorhinid from the Permian of Russia.] *Doklady Akad. Nauk URSS* 112:523–26. (In Russian.)

Watson, D. M. S. 1917. A sketch classification of the pre-Jurassic tetrapod vertebrates. *Proc. Zool. Soc. London* 1917:167–86.

———. 1921. On *Eugyrinus wildi* (A.S.W.), a branchiosaur from the Lancashire Coal Measures. *Geol. Mag.* 58:70–74.

———. 1940. The origin of frogs. *Trans. Roy. Soc. Edinburgh* 60:195–231.

———. 1961. Some additions to our knowledge of antiarchs. *Paleontology* 4:210–20.

Weber, M. 1904. *Die Säugetiere*. Jena. Pp. 866.

Weiner, J. S. 1955. *The Piltdown Forgery*. London. Pp. 214.

Welles, S. P. 1962. A new species of elasmosaur from the Aptian of Colombia and a review of the Cretaceous plesiosaurs. *Univ. Calif. Publ. Geol. Sci.* 44:1–96.

Westoll, T. S. 1938. Ancestry of the tetrapods. *Nature* (*London*) 141:127–28.

———. 1942*a*. Ancestry of captorhinomorph reptiles. *Nature* (*London*) 149:667–68.

———. 1942*b*. Relationships of some primitive tetrapods. *Nature* (*London*) 150:121.

———. 1958. The lateral fin-fold theory and the pectoral fins of ostracoderms and early fishes. *In*: *Studies on Fossil Vertebrates* (ed. T. S. Westoll). London. Pp. 180–211.

Westoll, T. S. and Miles, R. S. 1963. On an arctolepid fish from Gemünden. *Trans. Roy. Soc. Edinburgh* 65:139–53.

Westphal, F. 1963. Phytosaurier-Gattungen und -Arten aus dem südwestdeutschen Keuper (Reptilia, Thecodontia). *Neues Jahrb. Geol. Paläont. Abhandl.* 118(2):159–76.

Wetmore, A. 1956. A check-list of the fossil and prehistoric birds of North America and the West Indies. *Smithson. Misc. Coll.* 131(5): 1–105.

White, E. I. 1946*a*. The genus *Phialaspis* and the "*Psammosteus* Limestones." *Quart. Jour. Geol. Soc. London* 101:207–42.

———. 1946*b*. *Jamoytius kerwoodi*, a new chordate from the Silurian of Lanarkshire. *Geol. Mag.* 83:89–97.

———. 1950. The vertebrate faunas of the Lower Old Red Sandstone of the Welsh borders. *Bull. Brit. Mus.* (*Nat. Hist.*) 1(3):51–67.

———. 1958*a*. On *Cephalaspis lyelli* Agassiz. *Palaeontology* 1:99–105.

White, E. I. 1958*b*. Original environment of the craniates. *In*: *Studies on Fossil Vertebrates* (ed. T. S. Westoll). London. Pp. 212–34.

———. 1961. The Old Red Sandstone of Brown Clee Hill and the adjacent area. Part II. Palaeontology. *Bull. Brit. Mus.* (*Nat. Hist.*) *Geol.* 5(7):243–310.

———. 1962. A dipnoan from the Assise de Mazy of Hingeon. *Bull. Inst. Roy. Sci. Nat. Belg.* 38(50):1–7.

———. 1963. Notes on *Pteraspis mitchelli* and its associated fauna. *Trans. Edinburgh Geol. Soc.* 19:306–22.

———. 1966. Presidential address: a little on lung-fishes. *Proc. Linn. Soc. London* 177:1–10.

Whitworth, T. 1954. The Miocene hyracoids of East Africa. *Fossil Mammals of Africa No.* 7. *Brit. Mus.* (*Nat. Hist.*). London. Pp. 58.

Williston, S. W. 1914. The osteology of some American Permian vertebrates. *Contrib. Walker Mus.* 1:107–92.

Wilson, J. A. 1966. A new primate from the earliest Oligocene, West Texas, preliminary report. *Fol. Primat.* 4:227–48.

Wintrebert, P. 1922. La voûte palatine de *Lysorophus*. *C.R. et Mém. Soc. Biol. Paris* 87(2):928–30.

Wood, A. E. 1955. A revised classification of the rodents. *Jour. Mamm.* 36:165–87.

Wood, H. E. 1924. The position of the "sparassodonts": with notes on the relationships and history of the Marsupialia. *Bull. Amer. Mus. Nat. Hist.* 51:77–101.

Wood, R. C. 1967. A review of the Clark Fork vertebrate fauna. *Breviora, Mus. Comp. Zool.* No. 257:1–30.

Woods, J. T. 1956. The skull of *Thylacoleo carnifex*. *Mem. Queensland Mus.* 13:125–30.

———. 1962. Fossil marsupials and Cainozoic continental stratigraphy in Australia: a review. *Mem. Queensland Mus.* 14:41–49.

Woodward, A. S. 1921. Observations on some extinct elasmobranch fishes. *Proc. Linn. Soc. London* 133:29–39.

———. 1924. Un nouvel elasmobranche (*Cratoselache pruvosti* gen. et sp. nov.) du calcaire carbonifère inférieur de Denée. *Livre Jubilaire, Soc. Géol. Belg.* 1:59–62.

Young, C. C. 1947*a*. Mammal-like reptiles from Lufeng, Yunnan, China. *Proc. Zool. Soc. London* 117:537–97.

———. 1947*b*. On *Lufengosaurus magnus* Young (sp. nov.) and additional finds of *Lufengosaurus huenei* Young. *Paleont. Sinica*, new ser. C, No. 12:1–53.

———. 1948*a*. On two new saurischians from Lufeng, Yunnan. *Bull. Geol. Soc. China* 28:75–90.

———. 1948*b*. Fossil crocodiles in China with notes on dinosaurian remains associated with the Kansu crocodiles. *Bull. Geol. Soc. China* 28:255–88.

———. 1958*a*. On a new locality of *Yabeinosaurus tenuis* Endo and Shikama. *Vert. Palasiatica* 2:151–56.

———. 1958*b*. The first record of dinosaurian remains from Shansi. *Vert. Palasiatica* 2:231–36.

———. 1958*c*. The dinosaurian remains of Laiyang, Shantung. *Palaeont. Sinica*, new ser. C, No. 16:1–138.

———. 1958*d*. New sauropods from China. *Vert. Palasiatica* 2:1–28.

———. 1959*a*. On a new Stegosauria from Szechuan, China. *Vert. Palasiatica* 3:1–8.

———. 1959*b*. On a new Lacertilia from Chingning, Chekiang, China. *Sci. Rec. Peking*. n.s. 3:520–23.

———. 1964. The Pseudosuchians in China. *Palaeont. Sinica*, new ser. C, No. 19 (151):105–205.

Zangerl, R. 1966. A new shark of the family Edestidae, *Ornithoprion hertwigi*, from the Pennsylvanian Mecca and Logan Quarry shales of Indiana. *Fieldiana: Geology* 16(1):1–43.

Zangerl, R. and Richardson, E. S. 1963. The paleoecological history of two Pennsylvanian black shales. *Fieldiana: Geology Mem.* 4:1–352.

Zapfe, H. 1958. The skeleton of *Pliopithecus* (*Epipliopithecus*) *vindobonensis* Zapfe and Hürzeler. *Amer. Jour. Phys. Anthrop.* n.s., 16:441–58.

Indexes

Author Index

Subject Index